PHILOSOPHY OF SCIENCE

"A first-rate, challenging text that emphasizes the philosophy in the philosophy of science. Rosenberg offers a superb introduction to the epistemological and metaphysical issues at stake in modern science."

Professor Martin Curd, *Purdue University, Indiana*

"Philosophy students will like the way the issues in philosophy of science are connected to the basic concerns of epistemology and philosophy of language."

Professor Peter Kosso, *Northern Arizona University*

"An engaging and clearly written introduction to the philosophy of science. ... I was especially pleased to see the discussions of probability, the semantic view of theories, and science studies."

Peter Lipton, *Cambridge University*

Science is perhaps the sole distinctively western institution adopted by all cultures that have come in contact with it. And yet its scope, nature and methods have been contested throughout its history. Natural science has both raised the most fundamental questions for philosophers and shaped philosophers' theories about the nature of reality and the extent of our knowledge of it.

Philosophy of Science identifies the profound philosophical problems that science raises through an examination of enduring questions about its nature, methods and justification. Coming to grips with the nature of explanation, laws, causation, theory, models, evidence, reductionism, probability, teleology, realism and instrumentalism in science turns out to be a matter of facing the same questions that Plato, Aristotle, Descartes, Hume, Kant and their successors have grappled with.

This accessible and user-friendly text will be of value to students seeking an introduction to the philosophy of science as a central branch of philosophy and to students of science. It contains the following textbook features:

- chapter overviews and summaries
- a wide variety of clear supportive examples drawn from science
- study questions
- glossary and annotated further reading.

Alex Rosenberg is Professor of Philosophy at Duke University, Durham, North Carolina, USA. His ten books in the philosophy of science include *The Structure of Biological Science* (1985) and *Philosophy of Social Science* (1995). He has been a Guggenheim Fellow and a Fellow of the American Council of Learned Societies and the National Science Foundation. In 1993 he won the Lakatos Prize in the Philosophy of Science.

Routledge Contemporary Introductions to Philosophy

Series Editor:
Paul K. Moser
Loyola University of Chicago

This innovative, well-structured series is for students who have already done an introductory course in philosophy. Each book introduces a core general subject in contemporary philosophy and offers students an acessible but substantial transition from introductory to higher-level college work in that subject. The series is accessible to nonspecialists and each book clearly motivates and expounds the problems and positions introduced. An orientating chapter briefly introduces its topic and reminds readers of any crucial material they need to have retained from a typical introductory course. Considerable attention is given to explaining the central philosophical problems of a subject and the main competing solutions and arguments for those solutions. The primary aim is to educate students in the main problems, positions and arguments of contemporary philosophy rather than to convince students of a single position. The initial eight central books in the series are written by well experienced authors and teachers, and treat topics essential to a well-rounded philosophy curriculum.

Epistemology
Robert Audi

Ethics
Harry Gensler

Metaphysics
Michael J. Loux

Philosophy of Art
Noël Carroll

Philosophy of Language
William G. Lycan

Philosopy of Mind
John Heil

Philosophy of Religion
Keith E. Yandell

Philosophy of Science
Alex Rosenberg

PHILOSOPHY OF SCIENCE
A contemporary introduction

Alex Rosenberg

London and New York

First published 2000
by Routledge
11 New Fetter Lane, London EC4P 4EE

Simultaneously published in the USA and Canada
by Routledge
29 West 35th Street, New York, NY 10001

Routledge is an imprint of the Taylor & Francis Group

© 2000 Alexander Rosenberg

Typeset in Aldus Roman by Taylor & Francis Books Ltd
Printed and bound in Great Britain by Biddles Ltd, Guildford and King's Lynn

British Library Cataloguing in Publication Data
A catalogue record for this book is available from the British Library

Library of Congress Cataloging in Publication Data
Rosenberg, Alexander, 1946–
Philosophy of science: a contemporary introduction / Alex Rosenberg.
p.cm. – (Routledge contemporary introductions to philosophy)
Includes bibliographical references and index.
1. Science–Philosophy. I. Title. II. Series.

Q175 .R5475   2001
501–dc21   00–026960

ISBN 0–415–15280–1 (hb)
ISBN 0–415–15281–X (pb)

For
Eugene
and
Adrianne

Contents

Chapter 4: The structure and metaphysics of scientific theories 68

Chapter 5: The epistemology of scientific theorizing 106

Chapter 6: The challenge of history and post-positivism 135

Chapter 7: The nature of science and the fundamental questions of philosophy 158

Acknowledgments

This work began with the outrageous ambition of providing a worthy successor to Carl G. Hempel's splendid *Philosophy of Natural Science*, first published in 1966 and never yet improved upon. Browning's "Andrea Del Sarto" tells us, "Ah, but a man's reach should exceed one's grasp. Or what's a heaven for."

My second ambition is more attainable. It is to show that the problems of the philosophy of science are among the fundamental problems of philosophy, and that these problems emerge in our attempt to understand the nature of science.

My eventual recognition of this fact is the result of three decades of education by the likes of Peter Achinstein, Nick Rescher, Adolph Grunbaum, Richard Braithwaite, John Earman, David Hull, Michael Ruse, Bas van Fraassen, Elliot Sober, Philip Kitcher, Lindley Darden, Dan Hausman, Carl Hoefer, Marc Lange, Paul and Pat Churchland, Nancy Cartwright, Jarrett Leplin, Arthur Fine, Brian Skyrms, Paul Teller, Jan Cover, Paul Thompson, John Beatty, Ken Waters, Larry Hardin, Richard Boyd, Alison Wylie, Harold Kincaid, Steven Lukes, Richard Jeffrey, John Watkins, Alan Nelson, Tom Kuhn, Don Campbell, David Lewis, John Mackie, Wesley Salmon, Merrilee Salmon, Bill Newton-Smith, Joe Pitt, Robert Brandon, Larry Wright and Helen Longino.

In writing the present book, I have had detailed comments from Martin Curd, Peter Kosso, Peter Lipton, Neven Sesardic, Jarrett Leplin and Carl Hoefer, and extraordinary help from Marc Lange. I regret I could not produce a book that does full justice to all their advice and admonition.

This book would never have been started but for the research interests of Martha Reeves, and it would never have been completed but for the importunings of Moira Taylor.

CHAPTER 1
Why philosophy of science?

Overview

Philosophy of science is a difficult subject to define, in large part because philosophy is difficult to define. But on at least one controversial definition of philosophy, the relation between the sciences – physical, biological, social and behavioral – and philosophy are so close that philosophy of science must be a central concern of both philosophers and scientists. On this definition, philosophy deals initially with the questions which the sciences cannot yet or perhaps can never answer, and with the further questions of why the sciences cannot answer these questions.

Whether there are any such initial questions is itself a matter that can only be settled by philosophical argument. Moreover, if there are none, how science should proceed in its attempts to answer its as-yet-unanswered questions is also a matter for philosophical debate. This makes philosophy unavoidable for scientists. A cursory study of the history of science from the Greeks through Newton and Darwin to the twentieth century reveals these (as yet) scientifically unanswered questions.

Reflection on the way contemporary scientific findings and theories influence philosophy shows each is indispensable for understanding the other. Indeed, this chapter claims, and subsequent chapters argue, that philosophy is a fundamental prerequisite for understanding the history, sociology and other studies of science, its methods, achievements and prospects. Classical philosophical problems like those of free will versus determinism, or whether the mind is a part of the body, or whether there is room for purpose, intelligence and meaning in a purely material universe, are made urgent by and shaped by scientific discoveries and theories.

Science as a distinctive enterprise is arguably the unique contribution of western thought to all the world's other cultures which it has touched. As such, understanding science is crucial to our understanding of our civilization as a whole.

1 The relationship between science and philosophy

The history of science from the Greeks to the present is the history of one compartment of philosophy after another breaking away from philosophy and emerging as a separate discipline. Thus, by the third century BC, Euclid's work had made geometry a "science of space" separate from but still taught by philosophers in Plato's Academy. Galileo, Kepler and finally Newton's revolution in the seventeenth century made physics a subject separate from **metaphysics**. To this day, the name of some departments in which physics is studied is "natural philosophy". In 1859 *On The Origin of Species* set biology apart from philosophy (and theology), and at the turn of the twentieth century, psychology broke free from philosophy as a separate discipline. In the last fifty years, philosophy's millennium-long concern with logic has given rise to computer science.

But each of these disciplines which have spun off from philosophy have left to philosophy a set of distinctive problems: issues they cannot resolve, but must leave either permanently or at least temporarily for philosophy to deal with. For example, mathematics deals with numbers, but it cannot answer the question what a number is. Note that this is not the question what "2" or "dos" or "II" or "$10_{(\text{base 2})}$" is. Each of these is a numeral, an inscription, a bit of writing, and they all name the same thing: the number 2. When we ask what a number is, our question is not about the symbol (written or spoken), but apparently about the thing. Philosophers have been offering different answers to this question at least since Plato held that numbers were things, albeit abstract things. By contrast with Plato, other philosophers have held that mathematical truths are not about abstract entities and relations between them, but are made true by facts about concrete things in the universe, and reflect the uses to which we put mathematical expressions. But no one yet supposes that this approach does justice to the nature of mathematics.

Take another example, Newton's second law tells us that $\mathbf{F} = \mathbf{ma}$, force equals the product of mass and acceleration. Acceleration in turn is $\mathbf{dv/dt}$, the first derivative of velocity with respect to time. But what is time? Here is a concept we all understand, and one which physics requires. Yet both ordinary people, who surely know what time is, and physicists, for whom the concept is indispensable, would be hard pressed to tell us what exactly time is, or give a definition of it. Notice that to define time in terms of hours, minutes and seconds is to mistake the units of time for what they measure. It would be like defining space in terms of meters or inches. On the other hand, we can't say that time is duration, because duration is just the passage of, well, … time. Our definition would presuppose the very notion we set out to define. Explaining exactly what time is or defining it is a problem which science left to philosophy at least 300 years ago. With the advent of the general **theory** of relativity, physicists may well be taking back this question and finally addressing it themselves.

Similarly, many biologists and not a few philosophers have held that after Darwin, evolutionary biology took back from philosophy the problem of identifying the nature of man or the purpose or meaning of life. And some hold that what it shows is that man's nature is different only by degrees from that of other animals, and that there is no purpose and meaning to life. It is for this reason that evolutionary theory is so widely resisted; it purports to answer questions which should be left to philosophy.

All of the sciences, and especially the quantitative ones, depend heavily on the reliability of logical reasoning and **deductively valid arguments**; the sciences also rely on **inductive arguments** – ones which move from finite bodies of data to general theories. But none of the sciences address directly the question of why arguments of the first kind are always reliable, and why we should employ arguments of the second kind in spite of the fact that they are not always reliable. These are matters with which the subdiscipline of philosophy called logic broadly concerns itself.

What the history of science and the legacy of problems it leaves to philosophy shows is that the two intellectual inquiries have always been inextricably linked. And the legacy may help us define philosophy. One of the oddities about philosophy is that it seems to be a heterogeneous subject without the unity that characterizes, say, economics or chemistry. Among its subdisciplines, there is logic, the study of valid forms of reasoning; aesthetics, the study of the nature of beauty; ethics and political philosophy, which concern themselves with the basis of moral value and justice; **epistemology**, the study of the nature, extent and justification of knowledge; and metaphysics, which seeks to identify the fundamental kinds of things that really exist. What brings all these diverse questions together in one discipline? Here is a working definition of philosophy that identifies something these subdisciplines all have in common:

Philosophy deals with two sets of questions:

First, the questions that science – physical, biological, social, behavioral cannot answer now and perhaps may never be able to answer.

Second, the questions about why the sciences cannot answer the first lot of questions.

Some things to note about this working definition.

One type of question that only philosophy deals with is the **normative** questions – issues of value, questions about what ought to be the case, what we should do, about what is good and bad, right and wrong, just and unjust – in ethics, aesthetics and political philosophy. The sciences are presumably descriptive or, as is sometimes said, positive, not normative. Many of these normative questions have close cousins in the sciences. Thus, psychology will interest itself in why individuals hold some actions to be right and others wrong, anthropology will consider the sources of differences among cultures about what is good and bad, political scientists may study the consequences of various policies established in the name of justice, economics will consider how to maximize welfare subject to the normative assumption that welfare is what we ought to maximize. But the sciences – social or natural – do not challenge or defend the normative views we may hold. This is a task for philosophy.

In considering our working definition of philosophy, suppose one holds that in fact there are no questions that the sciences cannot now or cannot ever answer. One might claim that any question which is forever unanswerable is really a pseudo-question, a bit of meaningless noise masquerading as a legitimate question, like the question "Do green ideas sleep furiously?" or "When it's noon at Greenwich, what time is it on the sun?" Scientists and others impatient with the apparently endless pursuit of philosophical questions that seems to eventuate in no settled answers, may hold this view. They may grant that there are questions the sciences cannot yet answer, such as "What was happening before the big bang that began the universe?" or "How did inorganic molecules give rise to life?" or "Is consciousness merely a brain-process?" But, they hold, given enough time and money, enough theoretical genius and experimentation, all these questions can be answered, and the only ones left unanswered, at the end of scientific inquiry, will be pseudo-questions intellectually responsible persons need not concern themselves with. Of course, sapient creatures like us may not be around long enough in the history of the universe to complete science, but that is no reason to conclude that science and its methods cannot in principle answer all meaningful questions.

The claim that it can do so, however, needs an argument, or evidence. The fact that there are questions like "What is a number?" or "What is time?" that have been with us, unanswered, for centuries, is surely some evidence that serious questions may remain permanently unanswered by science. Could these really be pseudo-questions? We should only accept such a conclusion on the basis of an argument or a good reason. Suppose one wanted to argue that any question still left over at the "end of inquiry", when all the facts that science should attend to are in, must be pseudo-questions. As a philosopher I can think of some arguments in favor of this conclusion. But these arguments, that I for one can think of, all have two related features: first, they draw substantially on an understanding of the nature of science itself which science does not provide; second, these arguments are not ones science can construct by itself, they are philosophical arguments. And this is because they invoke normative premises, and not just the factual ones that science could provide. For example, the argument trades on the assumption that there are some considerations science *should, ought, is obliged to attend to*, as opposed to some things it can safely ignore. Which are the factors that make for knowledge that science should factor-in when deciding which questions are answerable, and what the answers to these questions are, and which are not answerable? This is a matter for epistemology – the study of the nature, extent and justification of knowledge. And this means that philosophy is unavoidable, even in the argument that there are no questions science cannot answer, either now or eventually or perhaps just "in principle".

Notice that this is not the conclusion that philosophers have some sort of special standing or perspective from which to ask and answer a range of questions that scientists cannot consider. These questions about science, its scope and limits are as much questions that scientists can contribute to answering as they are questions for philosophers. Indeed, in many cases, as we shall see, either scientists are better placed to answer these questions, or the theories and findings they have uncovered have an essential role in answering the questions. But the conclusion here is that philosophy is inescapable, even by those who hold that in the end all real questions, all questions worth answering, can only be answered by science. Only a philosophical argument can underwrite this claim.

2 Scientific questions and questions about science

Besides the questions science cannot answer yet, there are questions about why the sciences cannot yet or perhaps will not ever be able to answer these questions. Call the questions, about what a number is, or what time is, or what justice and beauty are, first-order questions. The second-order questions, about why science cannot as yet cope with the first-order questions, are themselves questions about what the limits of science are, how it does work, how it is supposed to work, what its methods are, where they are applicable and where not. Answering these questions will either enable us to begin to make progress on the hitherto unanswered first-order questions, or enable us to recognize that some of these first-order questions are not ones science can or needs to answer. Answering questions about what the nature of science and its method are can also help us assess the adequacy of proposed answers to scientific questions.

But there are other concerns – not directly scientific ones – in which the philosophy of science may be able to help us. Here are some important examples:

Philosophers, scientists and other defenders of the integrity of science and of its uniqueness as an instrument for the acquisition of objective knowledge have long opposed granting equivalent standing to non-scientific ways of belief-formation. They have sought to stigmatize astrology, crystal or pyramid power, or for that matter any New Age fashion, eastern mysticism, holistic metaphysics, as pseudo-science, as distractions, diversions, and unworthy substitutes for real scientific explanation and its application in practical amelioration of human life.

The issue is not purely academic. In the United States some years ago, an alliance was formed among groups of people impatient with the slow progress of orthodox empirical, experimental laboratory-based science to understand and deal with illness, together with those convinced that there was important therapeutically useful knowledge about illness, its causes and its cures, embedded in one or another non-experimental approach. This alliance prevailed upon the US Congress to direct the experimentally oriented National Institute of Health to establish an Office of Alternative Medicine mandated to spend significant sums of money (allegedly diverted from the funding of mainstream orthodox scientific research) in the search for such knowledge.

It is obviously difficult for opponents of this diversion of scarce resources from science, in support of what they consider wishful thinking and charlatanism, to argue that alternative medicine cannot provide knowledge, unless they have an account of what makes scientific findings into real knowledge.

On the other hand, advocates of such novel approaches have an equal interest in showing that it is in the nature of the orthodox scientific method to be blind to such non-experimental knowledge. Such advocates can make common cause with others – humanists, for example – who oppose what they call "scientism", the unwarranted overconfidence in the established methods of science to deal with all questions, and the tendency to displace other "ways of knowing" even in domains where conventional scientific approaches are inappropriate, unavailing or destructive of other goals, values and insights.

Both parties to this dispute have an equal interest in understanding the nature of science, both its substantive content and the methods by which it proceeds in the collection of evidence, the provision of explanations and the appraisal of theories. In other words, both sides of the debate need the philosophy of science.

Those who appreciate the power and the successes of the natural sciences, and who wish to apply methods successful in these disciplines beyond them to the social and behavioral sciences, have a special incentive to analyze the methods that have enabled natural science to attain its successes. Since the emergence of the social and behavioral sciences as self-consciously "scientific" enterprises, social and behavioral scientists, and some philosophers of science, have held that the relative lack of success of these disciplines, by contrast to the natural sciences, is due to a failure correctly to identify or implement the methods of natural science. For these students of social science, the philosophy of science has an obviously *prescriptive* role. Once it reveals the features of evidence-gathering, the explanatory strategies and the ways in which both are applied in technological advance by the natural sciences, the key to similar advance in the social and behavioral sciences becomes available. All the social and behavioral sciences need to do is employ the right method. Or so these students of scientific methodology argue.

And again, as with the controversy surrounding "alternative medicine", there are opponents of the scientific treatment of social and behavior issues. They wish to argue that the methods of natural science are inapplicable to their subjects, that "scientistic imperialism" is both intellectually unwarranted and likely to do harm by dehumanizing personal relationships and fragile social institutions. They go on to hold that such an approach is likely to be misapplied to underwrite morally dangerous policies and programs (for example, various eugenic policies pursued by many countries during the twentieth century), or even to motivate inquiry into areas best left unexamined (such as the genetics of behavior). It is clear that these defenders of the insulation of human affairs from scientific inquiry need both to understand what that inquiry consists in, and to identify those features of human conduct (for example "free will") which exempt it from scientific inquiry.

3 Modern science as philosophy

Besides the traditional questions which each of the sciences left as an intellectual legacy to philosophy, the development of the sciences over two millennia and more, has persistently raised new questions with which philosophers have struggled. Moreover, these two millennia of scientific development have shaped and changed the agenda of philosophical inquiry as well. Science has surely been the most powerful source of philosophical inspiration since its revolutionary successes of the seventeenth century.

Newton showed that motion – whether of planets and comets, or cannon balls and tides – was governed by a small number of simple, mathematically expressible and perfectly exceptionless laws. These laws were so exceptionless that given the position of the planets at any time at all, the physicist could calculate their position at any past time and any future time. If Newton is right, position and momentum at any one time fixes position and momentum for all future times. What is more, the same inexorable laws bind all matter, anything with mass. The determinism of Newtonian mechanics raised the specter of determinism in human behavior as well. For if humans are nothing but complex collections of molecules, i.e. of matter, and if these collections behave in accordance with the selfsame laws, then there is no real freedom of choice, there is only the illusion of it. Suppose we trace the causes of our apparently free actions, for which we are responsible, back through their previous causes to our choices, our desires and the physical states of our brains in which these desires are represented. If the brain is nothing but a complex physical object whose states are as much governed by physical laws as any other physical object, then what goes on in our heads is as fixed and determined by prior events as what goes on when one domino topples another in a long row of them. If the causes which fixed the events in our brain include events over which we have no control – our upbringing, our present sensory stimulation and physiological states, the environment, our heredity, then it may be claimed that there is no scope in this vast causal network for real free choice, for action (as opposed to mere behavior), and so none for moral responsibility. What is determined by the prior state of things and therefore beyond our control is not something for which we can be blamed – or praised for that matter.

With the success of Newton, determinism became a live philosophical option. But it remained open to some philosophers, and of course to many theologians, to hold that physics does not bind human action, or for that matter the behavior of any living thing. They held that the realm of the biological was beyond the reach of Newtonian determinism. And the proof of this was the fact that physical science could not explain biological processes at all, let alone with the power and precision that it explained the behavior of mere matter in motion.

Until the middle of the nineteenth century, opponents of determinism might comfort themselves with the thought that human action, and the behavior of living things generally, was exempt from the writ of Newtonian laws of motion. Human action and biological processes are evidently goal-directed, they happen for a purpose and reflect the existence of predestined ends which we strive to achieve and the vast scheme of things which God effortlessly attains. The biological realm shows too much complexity, diversity and adaptation to be the product of mere matter in motion; its appearance of design shows the hand of God. Indeed, before Darwin, the diversity, complexity and adaptation of the biological realm was the best theological argument for God's existence and for the existence of a "plan" that gives the universe meaning. This plan (of God's) was also at the same time the best scientific explanation for these three features of the biological realm. It was Darwin's achievement, as the theologians who opposed him so quickly realized and so strenuously denounced, to destroy the grounds of this theologically inspired metaphysical world-view. As Darwin wrote in his unpublished notebooks twenty years before he dared to publish the *On the Origin of Species*, "Origins of Man now proved. Metaphysics must flourish. He who understands baboon would do more towards metaphysics than Locke." I cannot summarize Darwin's alternative to revealed religion here (but see Chapter 3). Suffice it to say that if Darwin's evolutionary account of diversity, complexity and adaptation as the result of heritable genetic variation and natural environmental selection is right, there is no scope for a universe with meaning, purpose or intelligibility beyond the sort of clockwork determinism which Newton achieves. And this is a profoundly philosophical conclusion, which goes even beyond mere determinism by showing all purpose in nature to be illusory. Between them, Newton and Darwin are the great sources of philosophical materialism or physicalism, which undermines so much traditional philosophical theory in metaphysics and the philosophy of mind, and which for that matter may threaten moral philosophy.

However, twentieth-century developments in physics and the foundations of mathematics have shaken the confidence of philosophical materialism far more than any merely philosophical arguments. First, the attempt to extend deterministic physical theory from observable phenomena to unobservable processes came up against the appearance of subatomic indeterminism in nature. It has turned out that at the level of quantum processes – the behavior of electrons, protons, neutrons, the photons of which light is composed, and alpha, beta and gamma radiation – there are no exceptionless laws, the laws seem to be ineliminably probabilistic. It is not just that we cannot know what is going on with certainty and have to satisfy ourselves with mere probability. Rather, almost all physicists believe it has been physically established that the probabilities of quantum mechanics couldn't explain the behavior of the fundamental constituents of matter (and so of everything), with the fantastic precision that they reflect, if there were a deeper deterministic theory that somehow explains these probabilities. Whether a single particular uranium atom will emit an alpha particle in the next minute has a probability of, say, $.5 \times 10^{-9}$. No amount of further inquiry will raise or lower that probability; there is no difference in the state of a uranium atom which results in alpha emission during one minute and in the state of the atom when it does not emit the particle during the course of another minute. At the fundamental level of nature, the principle of same cause, same effect, is invariably violated.

Of course, by the time electrons, protons and other particles get lumped together into molecules, their behavior begins asymptotically to approach that of the determinism Newtonian mechanics demands. But Newton turns out to have been wrong, and in case one might hold out the hope that the world of observable objects Newton's theory deals with is exempt from quantum mechanical indeterminism, just recall that Geiger counters are observable detection devices whose clicking noises when held over radioactive materials enable quantum indetermined emissions of alpha particles to make an observably detectable difference in the macro-world.

Now, does all this mean that if determinism is false, free will and moral responsibility are after all vindicated as acceptable components of our philosophical world-view? Things are not that simple. For if the fundamental subatomic interactions that constitute our brain processes are not determined by anything at all, as quantum physics tells us, then there is even less room for moral responsibility in our actions. For actions will then stem from events that have no causes themselves, no reason at all for their occurrence. In short, quantum indeterminacy deepens the mystery of how human agency, deliberation, real choice, free will and ultimately moral responsibility is possible. Suppose that we can trace your actions, both the morally permissible and impermissible ones, back to an event, say, in your brain, which itself had no cause, but was completely random, undetermined and inexplicable, an event over which neither you nor anyone else nor for that matter anything else had any control whatsoever. Well, in that case, no one can be morally responsible for the effects of that event, including its effects in and on your desires, your choices, your actions.

If the direction in which science carries philosophy is a one-way street towards physicalism, determinism, atheism and perhaps even nihilism, then the intellectual obligation of those who wrestle with philosophical questions would be unavoidable. We must understand the substantive claims of physical science, we must be well enough informed to interpret the significance of these claims for philosophical questions, and we must understand the strengths and limitations of science as a source for answers to these questions.

But in fact, the direction in which science seems to carry philosophy is by no means a one-way street towards physicalism, determinism, atheism and nihilism. Since the sixteenth century many philosophers and scientists have endorsed the arguments of the mathematician, physicist and philosopher René Descartes that the mind is distinct from the body or any part of the body, in particular the brain. Descartes' followers have never argued that the mind can exist without the brain, any more than human life can exist without oxygen. But they held that (just as life is not merely the presence of oxygen) the mind is not identical to the brain. The mind is a separate and distinct substance, a non-physical one, and therefore not subject to the laws which physical science can uncover. If the mind is indeed not a physical thing, this may exempt humans and human action from obeying the **natural laws** science uncovers, or even from scientific study itself. It may turn out that humans and human actions must be understood by methods completely different from those which characterize natural science. Or it may be that human affairs cannot be understood at all.

This view, that the mind is non-physical and beyond the reach of natural science, may be greeted with dismay and stigmatized as obscurantist, and an obstacle to intellectual progress. But calling it names will not refute the arguments Descartes and others advanced in its behalf. And the general weakness of those social sciences inspired by methods and theories of natural sciences should give some further pause to those who reject Descartes' arguments. Can it really be that the only obstacle in social science to the sort of predictive precision and explanatory power we have in natural science is the greater complexity of human behavior and its causes?

Among those who answer this question in the affirmative have been psychologists and others who have sought to understand the mind as a physical device along the lines of the computer. After all, the neural architecture of the brain is in important respects like that of a computer: it operates through electrical signals that switch nodes of a network to states of "on" or "off". Psychologists interested in understanding human cognition have sought to model it on computers of varying types, recognizing that the human brain is vastly more powerful than the most powerful supercomputer and uses computational programs quite different from those with which we program current computers. But, if the brain is a powerful computer, and the mind is the brain, then at least modeling cognition by developing simple programs that simulate aspects of it on computers less powerful than the brain, will show us something about the mind by means of observing the output of a computer for a given input.

At this point, some argue that the development of science raises obstacles to this "scientistically" inspired research program. What we know for sure about computers is that they operate by realizing software programs with certain mathematical features. In particular, the software makes a computer operate in accordance with a system of mathematical axioms that enable it to derive an indefinite number of differing theorems. As a

simple example, consider the arithmetical calculations a comp\
expected to make. It can multiply any two numbers whatever. Th_ only\
way it can do so in a finite amount of time is to be programmed not with
the correct answer to every multiplication problem – there are infinitely
many of them, but to be programmed with the rules of multiplication in
the form of an axiom of arithmetic. Of course, there are limitations on the
calculations a computer can actually carry out. Anyone who has played with
a calculator knows what some of them are. If it runs out of power, or if the
numbers to be multiplied have too many places for the read-out screen, or if
an illegal operation like dividing by zero is attempted, or if the machine is
ordered to calculate pi, then it will not give a unique complete right answer.
In this respect computers are like human calculators.

But in the 1930s an Austrian mathematician, named Kurt Gödel,
proved mathematically that computers are not like human calculators in a
critical way. And subsequently some philosophers and scientists have
argued that this result is an obstacle to a scientific understanding of
cognition and of the mind. What Gödel proved was this: Any **axiomatic
system** powerful enough to contain all the rules for arithmetic will have
as a theorem, that is, as an implied consequence or a deductively derived
conclusion, at least one statement that is self-contradictory, provably
wrong, necessarily false. And any axiomatic system that is guaranteed to
be completely free of such contradictions will be incomplete – that is, will
lack at least one provable truth in arithmetic. This means that since
computer programs are axiomatic systems, no computer can ever be
programmed to do all the calculations of arithmetic correctly; at least one
calculation will be wrong if it can do all the right ones. And on the other
hand, the only way to program the computer to never make a mistake is
to make it at the same time incapable of doing all the correct calculations.
No computer can implement a program that includes arithmetic and that
is both complete and consistent. But, apparently, this is not a limitation on
us. To begin with, we humans, or at least one of us, Dr Gödel, proved this
result. He was able to do so because, unlike computers, minds like ours
can identify the one inconsistent statement in one axiom system-program
that is complete, and the one statement which is unprovable in the other
axiom system-program that avoids the inconsistency. So, evidently we, or
our minds, or at least the rules of thought we employ, are not merely the
software implemented on the hardware (or wetware) of our brains. Since
this mathematical result reflects a limitation on any physical system, no
matter what material it is made from – silicon chips, vacuum tubes, cogs
and wheels or neurons and synapses – it is argued, the human mind
cannot be material at all. And, therefore, it is not subject to study by

means appropriate to the study of material objects, whether those means are to be found in physics, chemistry or biology.

Here then is a result of modern science (and mathematics) which tends to undercut the confidence of the purely scientific world-view as a philosophy. Readers should be warned that the conclusion drawn above from Gödel's "incompleteness" proof, as it has come to be known, is highly controversial and by no means shared. Indeed, I do not accept the proof as showing anything like the conclusion drawn above. But the point is that results in science like this one are of overwhelming importance to the traditional agenda of philosophy, even when, as in this case, they suggest limitations on the scientific world-view as a philosophy.

4 Understanding science and understanding western civilization

Whether we like it or not, the only universal contribution of European civilization to all the rest of the world may be natural science. It is arguably the only thing developed in Europe which every society, culture, region, nation, population and ethnicity that has learned about it has adopted from Europe. The art, music, literature, architecture, economic order, legal codes and ethical and political value systems of the west have by no means secured widespread acceptance. Indeed, once decolonialization set in, these "blessings" of European culture have more often than not been repudiated by non-Europeans. But not so science. And we need not say "western" science. For there is no other kind, nor did science really spring up independently elsewhere before, simultaneously, or after its emergence among the Greeks 2500 years ago. It is true that some technologies that facilitated western political, military and economic dominance over much of the rest of the world, like gunpowder, moveable type and pasta, originated elsewhere, principally in China. And several non-western civilizations kept substantial and detailed records of celestial phenomena. But technological progress and astronomical almanacs are not science; the predictive powers that accompanied these achievements were not harnessed to an institutional drive to explain and improve discursive rational understanding that is characteristic of western science from the ancient Greeks to medieval Islam to Renaissance Italy to the Protestant Reformation and twentieth-century secularism.

There is thus something both characteristic about the west that it gave rise to science and something about science's origin that is symptomatic of distinctive western cultural values. The features of western culture and science that make one the origin of the other have been subject to considerable discussion among students of both. There has been no agreement on what these features are. Whatever the connection is, it is by no means one which precludes, inhibits or even discourages non-western peoples from excelling in all the sciences. And the reason that these people and peoples have taken on western science lock, stock and laboratory, seems obvious and unarguable: science has had a long-running record of continual predictive success, especially as evinced in technological improvements that enhance our control of our environment.

If it is the continual improvement of predictive power and technological control which commends science to non-western peoples, then presumably it is what sold western peoples on science as well. But this does not explain why it originated with western civilization, when no other important cultural institution of the same magnitude – religion, art, political institutions – seems to have had a unique origin in one culture. Perhaps the answer to this question is beyond reach. But one necessary step to answering it is that of understanding what science is, how it works, and what its methods, foundations and presuppositions are.

These are tasks which the philosophy of science long ago set itself. How do we distinguish these tasks from the tasks of late twentieth-century disciplines such as the sociology, psychology, economics of science and other social and behavioral studies of science? These disciplines have burgeoned in the last three decades, and now include large numbers of students of science eager to enhance our understanding of science. The philosophy of science has some claim to priority over these disciplines in the quest for an understanding of science.

To begin with, these intellectual enterprises are themselves presumably scientific: to the extent possible, they hope to share the methods of science in their own inquiries into the social, psychological, economic and political characteristics of science. Until we are clear about what the methods of science are, these enterprises are at risk of frustration and failure in attempting to attain their scientific objectives. For they will be unclear about the means to reach their scientific goals. This does not mean that we cannot do science of any kind until we have established what exactly the methods of science are, and ascertained their justification. But it means we should scrutinize those sciences already widely recognized as successful in the pursuit of their objectives, in order to identify the methods likely to succeed in less well-developed sciences, such as the sociology or psychology of science.

I believe this scrutiny cannot be sociological, psychological, economic or political, at least not at the outset. For science as a product – the concepts, laws, theories, methods of experiment and observation – and science as an enterprise of scientists, does not reflect the operation of factors studied in disciplines like sociology or psychology, economics, politics or history. The considerations that appear to drive scientific discussion, debate, acceptance and rejection are notions of logical reasoning, evidence, testing, justification, explanation, with which philosophy has grappled since Plato. If in the end analysis of and reflection on these notions and how they operate in science cannot answer our questions about its character nor sanction its claims to provide objective knowledge that other enterprises fail to secure, then we may usefully turn to the social and behavioral studies of the nature of science for real elucidation of the value of this distinctive contribution of the west to world civilization. But first we have to wrestle with the philosophy of science.

Summary

Philosophy is a hard discipline to define precisely, but the heterogeneous issues with which it deals all share in common a relationship to science. This chapter defines philosophy as the discipline that deals with the questions which science cannot answer, and with questions about why the sciences cannot answer the first set of questions.

The special place of science as a source of objective knowledge raises questions about how it secures such knowledge and whether there are alternative sources or means of securing it. Because it has always provided an influential description of reality, science has historically been the force most influential on the shape of pressing philosophical problems. Indeed, some philosophical problems track changes in natural science. How philosophers think about the mind and its place in nature, free will versus determinism, the meaning of life, all are deeply affected by scientific developments. As science's descriptions of reality have changed over the centuries, so the nature of the philosophical problems has changed as well.

Since science is arguably the only distinctive feature of western civilization that all the rest of the world has taken up, understanding science is an important part of coming to grips with the influence – good and bad – which it has had on other cultures.

Questions

Answering the study questions at the end of each chapter does not simply require a recapitulation of information provided in the chapter. Rather, they raise fundamental questions about philosophical theories treated in the chapter, and identify controversial issues on which readers are invited to disagree with the author, bring up examples, arguments and other considerations on which the text is silent, and make up their own minds. Some of the questions raised at the end of each chapter are worth revisiting after reading subsequent chapters.

1 The chapter begins with a potentially controversial definition of philosophy. Provide an alternative definition for philosophy, which accounts for the unity of the disparate parts of the discipline: metaphysics, epistemology, logic, ethics and political philosophy, aesthetics, etc.
2 Defend or criticize: "The claim that science is a uniquely western contribution to the world is ethnocentric, uninformed and irrelevant to understanding its character."
3 "As an open-minded and objective inquiry into the nature of the world, science should welcome the sort of unorthodox research which an agency like the Office of Alternative Medicine is designed to encourage." Are there good grounds for this claim?
4 Given the amount of change in the scientific conception of the world over the centuries, does philosophy pay too much attention to its findings and theories in dealing with philosophical problems?
5 Does the philosophy of science's conception of the nature of science compete with the sociology of science's conception?

Further reading

Readers seeking an introduction to the history of science, and especially its history since the Renaissance, will profit from Herbert Butterfield, *The Origins of Modern Science*. Thomas Kuhn, *The Copernican Revolution*, provides an account of seventeenth-century science by the historian of science most influential in its philosophy. I. Bernard Cohen, *The Birth of a New Physics*, and Richard Westfall, *The Construction of Modern Science*, provide accounts of Newtonian mechanics and its emergence. James B. Conant, *Harvard Case Histories in the Experimental Sciences*, is another influential source for understanding the history of the physical sciences.

Hans Reichenbach, one of the most important twentieth-century philosophers of science traces the influence of science on philosophy in *The Rise of Scientific Philosophy*. A classical work in the history of scientific and philosophical ideas is E. A. Burtt, *The Metaphysical Foundations of Modern Physical Science*, first published in 1926.

Richard Dawkins, *The Blind Watchmaker*, is an excellent introduction to Darwinism and the theory of natural selection. It is no substitute for reading Charles Darwin, *On the Origin of Species*, itself. The best introduction to the mysteries of quantum theory for the non-specialist is Richard Feynman, *QED: The Strange Story of Light and Matter*, while E. Nagel, and J. R. Newman, *Gödel's Proof*, provides an accessible account of this central mathematical result.

Important works in sociology of science start with R. Merton, *The Sociology of Science*. Views about the relation of sociology and philosophy of science quite different from those advanced here can be found in D. Bloor, *Knowledge and Social Imagery*. B. Barnes, D. Bloor and J. Henry, *Scientific Knowledge: A Sociological Analysis*, offers a revision of his earlier strong opposition. A. Pickering, *Constructing Quarks*, applies a sociological analysis to account for scientific discovery. Steven Shapin, *The Scientific Revolution*, brings the history and the sociology of science together in a way that reflects current thinking about the history of science by sociologists.

CHAPTER 2
Explanation, causation and laws

Overview

S cience, like other human activities, is one response to our need to understand the world. The way it does so differs from possibly competing activities like religion, mythology or for that matter common sense. And it claims to provide objective explanations superior in respects we value to these alternatives. These claims have been controverted in recent decades and need to be justified.

Alternative approaches to how science explains reflect fundamental philosophical differences, that go back to Plato, between those who view scientific explanation, like mathematical proof, as something we discover and those who treat it as something humans construct. Logical positivists aimed to formulate an ideal standard of explanation for scientists to aspire to. Other philosophers sought to understand how the reasoning works in explanations that scientists actually give.

One starting point for understanding scientific explanation focuses on the role of laws of nature. **Scientific laws** have explanatory force presumably because they describe the way things have to be. But the way things have to be, the **necessity** of laws of nature, is very difficult to understand from the scientific point of view. For scientific observation and experiment never show how things have to be, only how things are.

Dissatisfaction with answers to this question shifted the focus of some philosophers of science away from laws as explanatory. This approach leads to a theory of explanations which focuses on how explanations answer people's questions, instead of what ingredients they must have to be scientific.

1 Why do we need a theory of scientific explanation?

Philosophy, said Aristotle, begins with wonder. And by philosophy Aristotle meant science. Aristotle was right. Science seeks explanations to satisfy the wonder. But so do other human enterprises. The difference between science and other enterprises that seek explanations of why things are the way they are can be found in the sorts of standards that science sets itself for what will count as an explanation, a good explanation, and a better explanation. The philosophy of science seeks to uncover those standards, and the other rules that govern "scientific methods". It does so in part by examining the sorts of explanations scientists advance, accept, criticize, improve and reject. But what scientists accept or not as explanations cannot be the sole source of standards for what scientific explanation should be. After all, scientists are not infallible in their explanatory judgments; what is more, scientists themselves disagree about the adequacy of particular explanations, and about what explanation in science is like overall. If the philosophy of science were just a matter of collating the decisions of scientists about what explanations are, it could not be a source for advice about how scientific explanation *should* proceed. Yet in fact, in many disciplines, especially the social and behavioral sciences, scientists turn to philosophy of science for "prescriptions" – rules about how explanations ought to proceed if they are going to be truly scientific.

If the philosophy of science is to do more than merely describe what some or even many scientists take to be scientific explanations – if it is to endorse one or another recipe for scientific explanation as correct – it will have to do more than merely report what scientists themselves think about the matter. In addition to learning what explanations scientists actually accept and reject, the philosophy of science will have to assess these choices against philosophical theories, especially theories in epistemology – the study of the nature, extent and justification of knowledge. But this means that the philosophy of science cannot escape the most central, distinctive and hardest questions that have vexed philosophers since the time of Socrates and Plato.

Questions about the nature, extent and justification of knowledge, and in particular scientific knowledge, dominated philosophy from at least the time of Descartes and Newton, both of them important philosophers as well as scientists. For much of the twentieth century, the dominant answer to this question among philosophers of science was **empiricism**: the thesis that knowledge is justified by experience, that therefore the truths of science are not necessary, but **contingent truths**, and that knowledge could not extend beyond the realm of experience. Basing itself on this epistemology, a school of philosophy of science sprang up mainly in central Europe between the two world wars which adopted the label "logical positivist", or "logical empiricist" as later members of this movement called themselves. **Logical positivism** attempted to develop a philosophy of science by combining the resources of modern mathematical logic with an empiricist epistemology and a close study of the methods employed in the natural sciences, especially the physical sciences. Although **logical empiricism**'s answers to the central questions of the philosophy of science have been eclipsed, the questions it raised remain the continuing agenda of the philosophy of science: what is an explanation, a scientific law, a theory? Exactly how does empirical evidence decide on or choose between competing hypotheses? If empirical evidence does not suffice to choose between theories, or cannot do so, what should?

Could these questions be avoided if the philosophy of science gave up any pretense to prescription, or if scientists – natural or social – decided to ignore or reject the prescriptions of philosophers about how acceptable explanations should proceed? In recent years, some natural and social scientists, along with some historians, sociologists and even some philosophers, have rejected both the claim that the methods of science are open to assessment from the standpoint of philosophy, and the notion that philosophy might dictate to any other discipline how it should proceed, in explanation or any other activity. This view is often associated with labels such as "postmodernism" or deconstruction. It is treated further in Chapters 6 and 7. These students of scientific practice reject the relevance of epistemology or indeed of almost any considerations not drawn from their own particular disciplines to guide the methods of those disciplines. On their view, good economic methodology is what the leading economists are prized for doing; sound methods in psychology are whatever gets published in the major psychology journals; if the explanations of evolutionary biology differ in logic or evidence from those of chemistry, this could only show that biology's methods differ from those of chemistry, not that they are inadequate.

This tactic will not absolve the scientists from the responsibility of making choices about what are the correct methods in their fields, nor will it make philosophical problems go away. It will simply substitute one set of epistemological theories for another, and will embrace the philosophical theory that among the differing disciplines which contribute to human knowledge, there are few if any common factors that entitle them all to count as knowledge. This is an epistemological thesis itself in need of argument – philosophical argument. This means that for the scientist, the philosophy of science is unavoidable. Willy-nilly scientists must take sides on problems that have haunted our civilization since science began, that is, since philosophy began.

2 Defining scientific explanation

Traditionally the philosophy of science has sought a definition for "scientific explanation", but not a dictionary definition. A dictionary definition merely reports how scientists and others actually use the words "scientific explanation". Traditional philosophy of science seeks a checklist of conditions that any scientific explanation should satisfy. When all are satisfied, the checklist guarantees the scientific adequacy of an explanation. In other words, the traditional approach seeks a set of conditions individually necessary and jointly sufficient for something's being a scientific explanation. This "explicit" definition, or, as it was sometimes called, this "**explication**" or "rational reconstruction" of the dictionary definition, would render the concept of scientific explanation precise and philosophically well-founded.

An explicit definition gives the necessary and sufficient conditions for a thing, event, state, process or property to be an instance of the term defined. For example: "triangle" is explicitly defined as "plane figure having three sides". Since the conditions are together sufficient, we know that everything which fulfills them is a Euclidean triangle; and since the conditions are individually necessary, we know that if just one is not satisfied by an item, it is not a Euclidean triangle. The beauty of such definitions is that they remove vagueness, and provide for maximally precise definitions.

An explicit definition or "explication" of the notion of a scientific explanation could serve the prescriptive task of a litmus test or yardstick for grading and improving explanations in the direction of increasing scientific adequacy. The demand that philosophical analysis result in such a precise and complete definition is in part a reflection of the influence of mathematical logic on the logical positivists and their immediate successors in the philosophy of science. For in mathematics concepts are introduced in just this way – by providing explicit definitions in terms of already understood previously introduced concepts. The advantage of such definitions is clarity: there will be no borderline cases and no unresolvable arguments about whether some proposed explanation is "scientific" or not. The disadvantage is that it is often impossible to give such a complete definition or "explication" for most concepts of interest.

Let's call the sentences in an explanation which do the explaining the "*explanans*" (plural "*explanantia*"), and those which report the event to be explained the "*explanandum*" (plural "*explananda*"). There are no convenient English single-word equivalents for these terms, and so they have become commonplace in philosophy. An examination of the kinds of explanations that almost all scientists find acceptable makes it pretty obvious that scientific *explanantia* usually contain laws: when the *explanandum* is a particular event, like the Chernobyl reactor accident or the appearance of Halley's comet in the night sky over western Europe in the fall of 1986, the *explanans* will also require some "initial" or **"boundary conditions"**. These will be a description of the relevant factors – say the position and momentum of Halley's comet the last time it was sighted, or the position of the control rods of the reactor just before it overheated – which together with the law result in the *explanandum*-event. In the case of the explanation of a general law, like the ideal-gas law, **PV = rT**, the *explanans* will not contain boundary or **initial conditions**. Rather it will contain other laws, which work together to explain why this one obtains.

Suppose we want to know why the sky is blue, a question people have asked since probably as far back as any question. Now this is a particular state of affairs at a particular place, the earth. The Martian sky presumably is reddish in hue. So, to explain why the sky on earth is blue we require some information about boundary conditions and one or more laws. The relevant boundary conditions include the fact that the earth's atmosphere is composed of molecules mainly of nitrogen and oxygen. It's a law that gas molecules scatter the light which strikes them in accordance with a mathematical equation first formulated by the British physicist Rayleigh. The amount of light of any wavelength scattered by a gas molecule depends on its "scattering coefficient" – $1/\lambda^4$ – one over its wavelength to the fourth power. Since the wavelength of blue light is 400 nanometers (another law), and the wavelength of other light is greater (for example, red light has a wavelength of 640 nanometers), the scattering coefficient for blue light is greater than for other light. Therefore, the molecules in the earth's atmosphere will scatter more blue light towards the ground than other colors, and the atmosphere will look blue. In a physics text this explanation is laid out in more detail, the relevant equations derived and the amounts of scatter calculated.

Examples from the social and behavioral sciences are easier to understand because they are less quantitative. But explanations in social science that everyone accepts are harder to come by in these disciplines because we have discovered few if any laws in these disciplines. Thus, some economists will explain why the rate of interest is always positive (a general "law") by deriving it from other general "laws", such as the "law" that, other things being equal, people prefer immediate and certain consumption to future and uncertain consumption. From this law it follows that to get people to defer consumption to the future, you have to pay them by promising that they will have more to consume later if they postpone consumption, and instead invest what they would have consumed, to produce more. The payment for postponed consumption is measured as the interest rate. As in physics, the explanation here proceeds by derivation, this time of a law (instead of a particular fact), from other laws. Here we don't need boundary conditions because we are not explaining a particular fact. But the explanation still employs laws, if, that is, these generalizations about people are indeed laws. Some economists reject this explanation for why interest rates are always positive. They hold that other factors besides preference for immediate consumption explain this generalization.

Why must a scientific explanation contain one or more laws? What is it about laws which is explanatory? One answer begins with the claim that scientific explanation is causal explanation. Scientists search for causes. They do so because science seeks explanations which also enable it to control and predict phenomena, and this is something only knowledge of causes can provide. If scientific explanation is causal explanation, then by a well-known philosophical theory of **causation**, it must explicitly contain or implicitly assume laws. The empiricist account of causation holds that the relation of cause and effect obtains only when one or more laws *subsume* the events so related – that is, cover them as cases or instances of the operation of the law. Thus, the initial or boundary conditions of the *explanans* cite the cause of the *explanandum* phenomena, which are the effects of the boundary conditions according to the law mentioned in the *explanans*.

Causation consists in law-governed sequence on the empiricist view because there is no other observationally detectable property common and distinctive of all causal sequences besides exemplifying general laws. When we examine a single causal sequence – say one billiard ball hitting another, and the subsequent motion of the second ball – there is nothing to be seen that is not also present in a purely coincidental sequence, like a soccer goalkeeper's wearing green gloves and her successfully blocking a shot. The difference between the billiard-ball sequence and the green goalie-glove sequence is that the former is an instance of an oft-repeated sequence, and the latter is not. The last time the goalie wore the green gloves she failed to stop the shot.

All causal sequences share one thing in common that is missing in all coincidental sequences: they are instances of – they instantiate – general laws. This philosophical theory, which owes its origins to the eighteenth-century empiricist philosopher David Hume, does not require, for every causal claim we make, that we already know the law or laws which connect the cause and effect. Children will explain, correctly we suppose, why the vase broke, by admitting that it was dropped (passive voice, silence on who dropped it), on a marble floor. We accept the statement as identifying the cause, even though neither the children nor we know the relevant laws. Hume's theory doesn't require that we do so. It only requires that there is a law or laws, already known or not yet discovered, which do so. The task of science is to uncover these laws, and to employ them in explanations of effects.

If scientific explanation is causal explanation, and causation is law-governed sequence, then it follows pretty directly that scientific explanations require laws. The trouble with this argument for the requirement that scientific explanations appeal to laws is that first, a few important types of scientific explanations don't cite causes, or don't do so in any obvious way. The ideal-gas law, for example, explains the temperature of a gas at equilibrium by appeal to its simultaneous pressure and the volume it takes up. But these can't be causes since all three – the temperature, the volume, the pressure – obtain at the same time. Moreover, the nature of causation has been controversial in philosophy for hundreds of years. There is by no means a consensus on Hume's claim that every causal sequence is causal just because it is law-governed. Many philosophers have held that causation is a much stronger relation between events than mere regular succession. Thus, the sound of thunder regularly succeeds the flash of lightning, but the latter is not its cause. Rather they are joint effects of a common cause, the electrical discharge from the cloud to the earth. Most philosophers have agreed that causes somehow make their effects come about necessarily and that mere regularity cannot express this necessity. The logical empiricists who first advanced an explicit account of scientific explanation wished strongly to avoid traditional controversies about the existence and nature of causal necessity. Such questions were deemed "metaphysical" in the pejorative sense that no scientific experiment could answer them, and that no answer to them could advance scientific understanding of the world. In addition, some among the logical empiricists held that the notion of causation was an obsolete anthropomorphic one, with misleading overtones of human agency, manipulation or power over things. Accordingly, these philosophers needed a different argument for the requirement that scientific explanations must contain laws in their *explanans*.

The argument logical empiricists advanced for the role of laws in explanations illuminates several aspects of their philosophy of science. To begin with, these philosophers sought a notion of scientific explanation that would constitute an objective relationship between *explanandum* and *explanans*, a relationship like the relation of mathematical proof, which obtains regardless of whether anyone recognizes that it does, a relationship which is sufficiently precise that we can determine whether it obtains or not without any doubt or borderline cases. Thus, the logical empiricists rejected the notion of scientific explanation as an attempt to allay curiosity or answer a question which might be put by an inquirer. It is relatively easy to "explain" complex physical processes to children by telling them stories that allay their curiosity. The subjective psychological relevance of the *explanans* to the *explananda* in such cases may be very great, but they do not constitute scientific explanations. The logical empiricists were not interested in examining how a scientific explanation might be better or worse, appropriate or inappropriate, given the beliefs and interests of someone who might ask for the explanation. The conception of explanation as an answer to someone's question is not one these philosophers sought to explicate. They sought an explication of the concept of explanation which would provide it with the sort of role in science which the notion of "proof" plays in mathematics. The problem of explanation for logical empiricists was to find some conditions of explanation which insure the objective relevance of the *explanans* to the *explanandum*. They needed a relationship which made explanatory relevance a matter of objective relations between statements and not the subjective beliefs about relevance of less-than-omniscient cognitive agents.

We do well to pause here and contrast two fundamentally different philosophies of science. Some philosophers seek an objective relation between *explanandum* and *explanans* because they hold that science is constituted by truths about the world which obtain independently of our recognition, and which we set out to uncover. Thus science is treated in the way Plato, and his followers down to the present, conceive of mathematics as the study of objective relations between abstract objects, which obtain regardless of whether we recognize them. This approach to science may be more intuitively plausible than mathematical Platonism if only because the entities science seeks to uncover are not abstract, like numbers, but concrete, like genes.

By contrast with Platonism about mathematics, there are those who hold that mathematical truths are not about abstract entities and relations between them, but are made true by facts about concrete things in the universe, and reflect the uses to which we put mathematical expressions. Similarly, there are those who hold that science needs to be treated not like an abstract relation between truths, but as a human institution, a set of beliefs, and methods which we use to get around efficiently in the world. On this view scientific laws do not have a life of their own independent of the humans who invent and employ them. One might even try to capture this difference between philosophies of science by reflecting on the distinction between discovery and invention: Platonist-inclined philosophers treat the claims of science as truths to be discovered. By contrast there are the philosophers who treat science as a human institution, something we or the great scientists among us have invented to organize our experiences and enhance our technological control of nature. Platonists will seek an account of scientific explanation that makes it an objective relation between facts and/or statements that we set out to discover, while others seek a notion of explanation as an essentially human activity. The philosophy of science from which the logical empiricist model of explanation emerges is one which treats science as an act of discovery, not invention. In Section 4 we explore this subjective/objective contrast further.

The objective relevance relation on which the logical empiricists hit is the requirement that the *explanans* give good grounds to have expected the *explanandum*-event to have happened. You may be surprised by this requirement. After all, when we ask for the explanation of an event, we already know that it has happened. But satisfying this requirement involves producing further information which, had we been in possession of it before the *explanandum*-event occurred, would have enabled us to expect it, to predict it. Now, what kind of information would enable us to satisfy this requirement? A law and a statement of boundary or initial conditions will enable us to fulfill this requirement if the law and the boundary conditions together logically imply the *explanandum*. The relation of logical implication has two important features. First, it is truth-preserving: if the premises of a deductively valid argument are true, then the conclusion must also be true; second, whether the premises of an argument logically imply the conclusion is an objective matter of fact which can in principle be decided mechanically (for example, by a computer). These features answer to the very demand the logical empiricist makes of an explication of the concept of scientific explanation.

This analysis of scientific explanation, associated most closely with Carl G. Hempel, the philosopher who did the most to expound and defend it, came to be called the "**deductive-nomological (D-N) model**" ("nomological" from the Greek *nomos*, meaning lawful). Critics of this D-N account of explanation labeled it (and its statistical extensions) the "**covering law model**", and this name came to be adopted by its defenders as well. Hempel's fundamental idea was the requirement mentioned above, that the *explanans* give good grounds to suppose that the *explanandum* phenomenon actually occurred. It stands as his "general adequacy criterion" on scientific explanations.

In Hempel's original version the requirements on deductive nomological explanation were as follows:

1 The explanation must be a valid deductive argument.
2 The *explanans* must contain at least one general law actually needed in the deduction.
3 The *explanans* must be empirically testable.
4 The sentences in the *explanans* must be true.

Between them, these four conditions are supposed to be individually necessary and jointly sufficient conditions for any set of statements to constitute a scientific explanation of a particular fact. Notice that an explanation which satisfies these conditions provides enough information so that one could have predicted the occurrence of the *explanandum-event*, or similar events, given that one knows that the initial or boundary conditions obtain. Thus, the D-N model is committed to the symmetry in principle of explanation and prediction. In fact, this commitment already follows from the objective relevance requirement stated above.

The first condition guarantees the relevance of the *explanans* to the *explanandum*. The second condition is so stated to exclude as an explanation a patently non-explanatory argument like:

1 All free-falling bodies have constant acceleration.
2 It rained on Monday.

Therefore,
3 It rained on Monday.

Notice this argument satisfies all the other conditions on explanation. In particular, it is a deductively valid argument because every proposition deductively implies itself, so 2 implies 3. But it is no explanation, if only because nothing can explain itself! And of course it's not a D-N explanation for another reason: the law it includes is not needed to make the deduction valid. Consider another example.

1 All puppies born in this litter have a brown spot on their foreheads.
2 Fido is a puppy born in this litter.

Therefore,
3 Fido has a brown spot on his forehead.

This argument is no explanation of its conclusion owing to the fact that premise 1 is no law of nature. It's an accident of genetic recombination at best.

The third condition, **testability**, is supposed to exclude non-scientific explanations that make reference to explanatory factors that cannot be subject to confirmation or disconfirmation by observation, experiment or other empirical data. It reflects the epistemological commitment of empiricism about scientific knowledge: the requirement that the *explanans* be testable is meant to exclude non-scientific and pseudo-scientific explanations, such as those offered by astrologers, for example. How testability is assured is a subject to which we turn in Chapter 5.

The fourth condition, that the *explanans* be true, is problematical and introduces some fundamental philosophical problems, indeed the very ones the logical empiricists hoped to escape by silence about causation. Every scientific explanation must include a law. But laws are by definition true everywhere and always, in the past, in the present, in the future, here and everywhere else in the universe. As such, they make claims that cannot be established conclusively. After all, right now we have no access to the distant past or even the nearest future, let alone all places and times where events happen that make laws true. That means that the statements we believe to be laws are at best hypotheses which we cannot know for sure to be true (see Section 4, below). For convenience let's distinguish between "natural laws", true everywhere and always whether we have uncovered them or not, and "scientific laws", which is what we will call those hypotheses well established in science as our best current estimates of what the natural laws are.

Since we cannot know whether our scientific laws are natural laws, that is, whether they are true, we cannot ever know for sure that any explanation satisfies condition 4 above: that the *explanans* be true. Indeed, the situation is worse: since every previous hypothesis we have advanced about the natural laws has proved to be wrong, and been replaced by a more accurate scientific law, we have excellent reason to suppose that our current scientific laws (our current best guesses about what the natural laws are) are wrong too. In that case, we have equally good reason to think that none of our current scientific explanations really satisfy the deductive nomological model. For we have reason to believe that at least one of their *explanantia* – the scientific law – is false!

But what's the use of an analysis of explanation according to which we probably have never uncovered any scientific explanations, only at most approximations to them, whose degree of approximation we can never measure?

We might try to avoid this problem by weakening requirement 4. Instead of requiring that the *explanans* be true, we might require that the *explanans* be true or our best current guesses about the natural laws. The trouble with this weakened requirement is twofold. It is by no means clear and precise which are our best guesses about natural laws. Physicists disagree just as social scientists do about which guess is the best one, and philosophers of science have by no means solved the problem of how to choose among competing hypotheses. In fact, the more one considers this question the more problematical becomes the nature of science, as we shall see in Chapters 3 and 4. Weakening the requirement of truth into the requirement that the *explanans* include the most well-established currently known scientific law (i.e. our best-guess hypothesis) thus undermines the D-N model's claims to precision in explication.

The second problem we face is the nature of scientific laws and natural ones. Two of our four conditions on a scientific explanation invoke the notion of a law. And it is pretty clear that the explanatory power of a scientific explanation is in fact borne by the law. This is something even those who reject the covering law model of explanation accept (as we shall see below). The scientific law is what makes the connection between the particular facts mentioned in the initial conditions of the *explanans*, and the particular facts mentioned in the *explanandum*. If we seek an account of what makes a D-N argument explanatory, the source must be at least in large part in the law it invokes. But what exactly is a natural law?

3 Why do laws explain?

The logical empiricists early identified several features of a law on which there has continued to be wide agreement: laws are universal statements of the form "all **A**s are **B**s" or "if event **E** happens, then invariably, event **F** occurs". For example, "all pure samples of iron conduct electric currents at standard temperature and pressure" or "if an electric current is applied to a sample of iron under standard temperature and pressure, then the sample conducts the current". These are terminological variants of the same law. Philosophers tend to prefer the "if ... then ... " conditional version to express their form. Laws don't refer to particular objects, places or times, implicitly or explicitly. But these two conditions are not suffi- cient to distinguish laws from other statements grammatically similar to laws but without explanatory force. Compare the two following state- ments of the same universal form:

> All solid spherical masses of pure plutonium weigh less than 100,000 kilograms.

> All solid spherical masses of pure gold weigh less than 100,000 kilograms.

We have good reason to believe that the first statement is true: quantities of plutonium spontaneously explode long before they reach this mass. Thermonuclear warheads rely on this fact. There is also good reason to think that the second statement is true. But it is true just as a matter of cosmic coincidence. There could have been such a quantity of gold so configured somewhere in the universe. Presumably the former statement reports a natural law, while the latter describes a mere fact about the universe that might have been otherwise. One way to see that the state- ment about plutonium is a law is that an explanation of why it is true requires us to appeal to several other laws but no initial or boundary conditions; by contrast, to explain why there are no solid gold spheres of 100,000 kilograms requires laws and a statement of boundary or initial conditions that describe the distribution of atoms of gold in the universe from which gold masses are formed. What this shows is that universality of form is not enough to make a statement a law of nature.

One symptom of the difference between real laws and accidental generalizations philosophers have hit upon involves grammatical constructions known as "**counterfactual conditionals**", or "counterfactuals" for short. A counterfactual is another sort of if/then statement, one expressed in the subjunctive tense, instead of the indicative tense in which laws are expressed: We employ such statements often in everyday life: "If I had known you were coming, I would have baked a cake." Two examples of such counterfactual statements relevant for distinguishing laws from non-laws of the same grammatical "if ... then ... " form are the following:

> If it were the case that the moon is made of pure plutonium, it would be the case that it weighs less than 100,000 kilograms.

> If it were the case that the moon is made of pure gold, it would be the case that it weighs less than 100,000 kilograms.

Notice that the antecedents (the sentences following the "ifs") and the consequents (the sentences following the "thens") of both counterfactuals are false. This grammatical feature of counterfactual sentences is obscured when we express them more colloquially and less stiltedly as follows:

> If the moon had been composed of pure plutonium, it would weigh less than 100,000 kilograms.

> If the moon had been composed of pure gold, it would weigh less than 100,000 kilograms.

So, these two statements are claims not about actualities, but about possibilities – the possible states of affairs that the moon is composed of plutonium and gold respectively. Each says that if the antecedent obtained (which it doesn't), the consequent would have obtained (even though, as a matter of fact, neither does actually obtain). Now, we hold that the counterfactual about gold is false. But we believe that the counterfactual about plutonium truly expresses a truth. And the reason for this difference between these two grammatically identical statements about non-actual states of affairs is that there is a law about plutonium that supports the plutonium counterfactual, while the universal truth about gold masses is not a law, but merely an accidental generalization. So, it does not support the gold counterfactual.

Thus, we may add to our conditions on laws that in addition to being universal in form, they support counterfactuals. But it is crucial to bear in mind that this is a symptom of their being laws, not an explanation of it. That is, we can tell the difference between those generalizations we treat as laws and those we do not by considering which counterfactuals we accept, and which we do not accept. But unless we understand what makes true counterfactuals true independent of the laws which support them, the fact that laws support counterfactuals won't help explain the difference between them and accidental generalizations.

We know that laws support their counterfactuals, while accidental generalizations do not. But we don't know what it is about laws that makes for this difference. Presumably, they support their counterfactuals because laws express some real connection between their antecedents and their consequents that is missing between the antecedent and the consequent of an accidental generalization. Thus, there is something about being a sphere of pure plutonium that brings it about, or necessitates the fact, that it cannot be 100,000 kilograms in mass, whereas there is nothing about being a sphere of gold that makes it impossible to be that massive.

But what could this real connection between the antecedent and the consequent of a law be, which reflects the necessitation of the latter by the former? Certainly, laws do not express logical necessities. Or at least this is widely believed in the philosophy of science on the ground that the denial of a natural law is not contradictory, whereas the denial of a logically necessary statement, like "all whole numbers are either odd or even" is contradictory. It's impossible to conceive of the violation of a logically **necessary truth**. It's easy to conceive of the violation of a natural law: there would be nothing contradictory about gravity varying as the cube of the distance between objects instead of as the square of the distance between them. Laws of nature cannot be logically necessary.

It's no explanation of the necessity of laws to say they reflect "nomological" or "physical" or "natural" instead of **logical necessity**. A statement is logically necessary if its denial is a self-contradiction or, equivalently, if its truth is required by the laws of logic. On this model, what is it for a statement to be of physical or natural necessity except that it is required to be the case by the laws of physics or nature? If this is what natural or **physical necessity** consists in, then grounding the necessity of laws on physical or natural necessity is grounding the necessity of laws on itself! This is reasoning in a circle, and it can lead nowhere.

This question of what kind of necessity laws have, and accidental generalizations lack, is exactly the sort of "metaphysical" question that the logical empiricists hoped to avoid by not invoking the notion of causality in their analysis of explanation. For nomological necessity just turns out to be the same thing as the necessity that connects causes and their effects and is missing in merely accidental sequences. The nature of the causal connection turns out to be unavoidable even if it is metaphysical. But perhaps we can make progress understanding what makes a generalization a law by thinking more about causality. At a minimum the connection between the necessity of laws and causation will illuminate the sense in which scientific explanation is causal even when the words cause and effect do not figure in the explanation.

Recall our discussion of causal sequences versus coincidences. Presumably a causal sequence is one in which the effect is brought about by the cause, produced by it, made to happen by the cause's occurrence, necessitated by it; one way of making this point is to put it this way: "if the cause hadn't happened, the effect would not have happened" – the counterfactual kind of statement we encountered when trying to understand the necessity of laws. By contrast to a causal sequence, there is no such relation of necessitation between the first event and the second in a coincidental sequence. But what does this causal necessitation consist in? There does not seem to be any "glue" or other observationally or theoretically detectable connection between events in the universe. All we ever see, even at the level of the microphysical, is one event, followed by another event. Try the thought experiment: consider what goes on when one billiard ball hits another one and the second moves; the transfer of momentum from the first to the second is just a way of saying that the first one moved, and then the second one did. After all, momentum is just (mass × velocity) and the masses didn't change, so the velocity must have changed when the momentum was transferred. Consider the counterfactual that "if the momentum hadn't been transferred to the second ball, it would not have moved". Why not? Will it help to consider what happened at the level of the molecules out of which the billiard balls are made? Well, the distance between them became smaller and smaller until suddenly it started to grow again as the balls separated. But there was nothing else that happened below the level of observation besides the motion of the molecules in the first billiard ball, followed by the motion of the molecules that made up the second. Nothing so to speak hopped off the first set of molecules and landed on the second set; the first set of molecules didn't have a set of hands which reached out and pushed the second set of molecules. And if we try the thought experiment at a deeper level, say the level of atoms, or the quarks and electrons that make up the atoms, we will still only see a sequence of events, one following the other, only this time the events are subatomic. In fact, the outer shell electrons of the molecules on the surface of the first ball don't even make contact with the electrons on the outer shells of the molecules at the nearest surface of the second ball. They come close and then "repel" each other, that is, move apart with increasing acceleration. There does not appear to be any glue or cement that holds causes and effects together that we can detect or even imagine.

If we cannot observe or detect or even conceive of what the necessary connection between individual instances of causes and their effects might be, the prospect for giving an account of how causal explanation works or why laws have explanatory force becomes dimmer. Or at least the logical empiricists' hope to do this in a way that avoids metaphysics will be hard to fulfill. For the difference between explanatory laws and accidental generalizations, and the difference between causal sequences and mere coincidences, appears to be some sort of necessity that the sciences themselves cannot uncover. If the question of why laws explain has been answered by the claim that they are causally or physically or nomologically necessary, the question of what causal or physical or nomological necessity is remains as yet unanswered. Answering the question takes us from the philosophy of science into the furthest reaches of metaphysics, and epistemology, where the correct answer may lie.

4 Counterexamples and the pragmatics of explanation

Progress in the philosophy of science has often consisted in the construction of **counterexamples** to analyses, definitions or explications, and then revisions of the definition to accommodate the counterexamples. Since the sort of analysis traditionally preferred by logical empiricists provides a definition in terms of conditions individually necessary and jointly sufficient for the concept being explicated, counterexamples can come in two different forms: first, examples that most informed persons will concede to be explanations but which fail to satisfy one or more of the conditions laid down; second, examples that no one takes to be acceptable scientific explanations, but which satisfy all conditions.

Counterexamples to the D-N model of the first sort have often been found in history and the social sciences, where the most well-accepted explanations often fail to satisfy more than one of the D-N model's conditions, especially the requirement that laws be cited. For example, the explanation of why Britain entered the First World War against Germany does not seem to involve any laws. Imagine someone framing a law of the form, "Whenever Belgian neutrality is protected by treaty and is violated, then the treaty signatories declare war on the violator." Even if the proposition is true, it's no law, not least because it names a specific place in the universe. If we substitute for "Belgian" something more general, such as "any nation's", the result is more general, but plainly false. One response to the fact that many explanations don't cite laws that is often made in defense of D-N explanation is to argue that such explanations are "explanation sketches" which could eventually be filled out to satisfy D-N strictures, especially once we have uncovered all the boundary conditions and the relevant laws of human action. Counterexamples of this sort in the natural sciences are more difficult to find, and defenders of the D-N model are confident they can deal with such cases by arguing that the alleged counterexample does satisfy all conditions. Thus, consider the explanation of the *Titanic's* sinking. Her sinking was caused by collision with an iceberg. Surely this explanation will be accepted even though there is no law about the *Titanic*, nor even one about ships that strike icebergs sinking. The explanation is an acceptable one even when we note that it is often offered and accepted by persons who know almost nothing about the tensile strength of iron, the coefficient of elasticity of ice, or the boundary conditions which obtained on the night of 12 April 1912 in the North Atlantic. Presumably, a naval engineer could cite the relevant laws along with the boundary conditions – size of the iceberg, speed of the *Titanic*, composition of its hull, placement of its watertight doors, etc. which underlie the explanation-sketch, and which enable us to turn it into a D-N explanation.

Counterexamples of the second sort, which challenge the sufficiency of the D-N conditions as a guarantee of explanatory adequacy, are more serious. Among the most well-known is the "flagpole's shadow counter-example" due to Bas van Fraassen. Consider the following "explanation" for the fact that at 3.00 p.m. on 4 July 2000, the flagpole at City Hall in Missoula, Montana, is 50 feet high:

1 Light travels in straight lines. (a law)
2 At 3.00 p.m. on 4 July 2000 the sun is emitting light at a 45 degree angle to the ground where the flagpole is located, perpendicular to the ground. (boundary condition)
3 The shadow cast by the flagpole is 50 feet long. (boundary condition)
4 A triangle with two equal angles is isosceles. (mathematical truth)

Therefore:
5 The flagpole is 50 feet high.

The "explanation" is designed to satisfy all four conditions given for D-N explanations above, without being a satisfactory explanation of the height of the flagpole. The deductive argument fails to be an explanation presumably because it cites an effect of the flagpole's height – the shadow it casts – and not its cause – the desires of the Missoula city mothers to have a flagpole one foot taller than the 49-foot flagpole at Helena, Montana.

One conclusion sometimes drawn from this counterexample is simply to reject the whole enterprise of seeking an objective explanatory relation between statements about facts in the world independent of the human contexts in which explanations are requested and provided. To see why such a move might be attractive, consider whether we could construct a context in which the deduction above is in fact an acceptable explanation for the height of the flagpole. For example, suppose that the city mothers had wished to build the flagpole to commemorate the American commitment to equality and union by casting a shadow exactly equal in length to the pole and exactly as many feet as there are states in the union at the moment annually chosen for patriotic exercises on American Independence Day. In that case, van Fraassen argued, for someone well-informed about the wishes of the city mothers, it would be a correct answer to the question "Why is the flagpole 50 feet high?" to reply in the terms mentioned in the deductive argument above.

This argument is supposed to show that explanation is not merely a matter of logic and meaning – syntax and semantics; it is as much a matter of "**pragmatics**" – that dimension of language which reflects the practical circumstances in which we put it to use. We may contrast three different aspects of a language: its syntax, which includes the rules of logic as well as grammar; its semantics – the meanings of its words; and its pragmatics, which includes the conditions that make some statements appropriate or meaningful. For example, it's a matter of the pragmatics of language that "Have you stopped beating your wife, answer yes or no?" is a question we can only ask wife-beaters. An unmarried man or one not given to wife-beating cannot answer this question with a yes or a no. Similarly, if explanation has a pragmatic element we can't tell when something successfully explains unless we understand the human context in which the explanation was offered.

The pragmatics of language is presumably something we can ignore in mathematical proof, but not, it is argued, in scientific explanation. Whether an analysis of scientific explanation must include this pragmatic dimension is a topic for the next chapter. But one point that can be made is that even if explanation is unavoidably pragmatic, it may still turn out that the D-N model provides important necessary conditions for scientific explanation – to which some pragmatic conditions need be added. Indeed, it may be that the D-N model provides the distinctive features of *scientific* explanation, while the pragmatic element provides the features common to scientific and non-scientific *explanations*.

Another implication sometimes drawn from the flagpole counterexample is that the D-N model is inadequate in not restricting scientific explanations to causal ones, or at least in not excluding from the *explanans* factors later in time than the *explanandum*. Notice that the casting of a shadow 50 feet long at 3.00p.m. on 4 July 2000 is something that happens well after the flagpole was first fabricated at 50 feet in height or mounted vertically. But what is the reason for this restriction?

Evidently it is our belief that causation works forward in time, or at least not backwards, and that somehow the direction of explanation must follow the direction of causation. So, we might add to the D-N model the additional condition that the boundary conditions be the prior causes of the *explanandum*. The trouble with this addition to our requirements on explanation is that there appear to be scientific explanations that do not invoke temporally prior causes. Suppose for example we explain the temperature of a gas at equilibrium in terms of the ideal-gas law, $PV = rT$, and the boundary condition of its simultaneous pressure and the volume of the vessel in which it is contained. If this is a causal explanation, it is not one which cites causes earlier in time.

Worse still, this addition invokes causation to preserve the D-N model, and causation is something about which the proponents of D-N explanation wanted to remain silent. Although the logical empiricists tried, philosophers of science were eventually unable to continue to maintain a dignified silence about the embarrassingly metaphysical problems of causation, owing to another obligation they bore: that of providing an account of how statistical explanation works. Both the social and biological sciences have long been limited to providing such explanations just because they have not uncovered universal non-statistical laws. And the indeterminacy of subatomic physics makes such explanations arguably unavoidable, no matter how much we learn about nature.

It may seem a straightforward matter to extend the D-N model to statistical explanations. But it turns out that the straightforward extension is another reason to take the pragmatics of explanation seriously, or at least to treat explanation as a relation between facts about the world and the beliefs of cognitive agents who ask for explanations.

For example, to explain why Ms R. votes for the left-of-center candidate in the latest election, one may cite the boundary condition that both her parents always did so, and the statistical law that 80 per cent of voters vote for candidates from the same location on the political spectrum as their parents voted for. The form of explanation is thus an argument with two premises, one of which is a general law, or at least an empirical generalization that is well supported.

Explanans:
1 80 per cent of voters vote for candidates from the same location on the political spectrum as the candidates that their parent of the same gender voted for. (well-confirmed statistical generalization)
2 Ms R.'s mother voted for left-of-center candidates. (boundary condition)

Therefore, with .8 probability,
Explanandum:

3 Ms R. voted for the left-of-center candidate in the latest election.

But clearly the argument form of this explanation is not deductive. The truth of the premises does not guarantee the truth of the conclusion: they are compatible with the women in question not voting at all, or voting for the right-of-center candidate, etc.

Statistical explanations on this view are inductive arguments – that is, they give good grounds for their conclusions without guaranteeing them, as deductive arguments do. It is no defect of inductive arguments that they are not truth-preserving, or do not provide guarantees for their conclusions (assuming the premises are true) the way deductive arguments do. All scientific reasoning from a finite body of evidence to general laws and theories is inductive – from the particular to the general, from the past to the future, from the immediate testimony of the senses to conclusions about the distant past, etc. (This is a matter on which we will focus in Chapter 5.)

In this case, the 80 per cent frequency of voters voting as did their same-gender parents may be held to provide an 80 per cent probability that Ms R. can be expected to vote as she did. Thus, like D-N explanations, a so-called **inductive-statistical (I-S) model of explanation** gives good grounds that the *explanandum* phenomenon can be expected to occur. However, there is a serious complication that the I-S model must deal with. Suppose that in addition to knowing that both Ms R.'s parents voted for candidates of the left, we also know that Ms R. is a self-made millionaire. And suppose further that we know that it is a statistical generalization that 90 per cent of millionaires vote for right-of-center candidates. If we know these further facts about Ms R. and about voting patterns, we can no longer accept as an explanation of why she voted left that her parents did and 80 per cent of voters vote as their parents did. For we know that it is 90 per cent probable that she voted for the right-of-center candidate. Evidently we need some other statistical or non-statistical generalization about female millionaires whose parents voted left to provide a statistical explanation for why Ms R. did so. Suppose that the narrowest class of voters studied by political scientists includes female self-made millionaires from Minnesota, and that among these 75 per cent vote for candidates of the left. Then we may be entitled to explain why Ms R. so voted by inductively inferring from this generalization and the fact that she is a self-made millionaire from Minnesota that she voted as she did, and this will count as an I-S explanation of the fact. It is because this is the narrowest class of voters about which we have knowledge, that we know which among these statistical regularities (all of them true) to apply in the explanation. So, to get an account of I-S explanation, we need to add to the four conditions on D-N explanation, something like the following additional condition:

5 The explanation must give a probability value for the conclusion no higher than the probability given in the narrowest relevant reference class the *explanandum* phenomenon is *believed* to fall into.

But notice, we have now surrendered a fundamental commitment of the logical empiricist's approach to explanation: we have made the subjective beliefs of agents who ask for and offer explanations an essential element in scientific explanation. For it is our beliefs about the narrowest relevant reference class for which we have framed statistical regularities that determines whether an explanation satisfies the requirements of the I-S model. Of course, we could drop the qualification "is believed to" from condition 5, but if the underlying process that our statistical generalization reports is really a deterministic one, our I-S explanation will reduce to a D-N model, and we will have no account of statistical explanation at all.

Perhaps the problems of statistical explanation and the flagpole's shadow counterexample should lead us to take seriously alternatives to the logical empiricist theory of explanation that emphasize the epistemic and pragmatic dimensions of explanation. Instead of starting with a strong philosophical theory and forcing scientific practice into its mold, these approaches are sometimes claimed to take more seriously what scientists and others actually seek and find satisfactory in explanations.

One way to see the differences between the pragmatic/epistemic approach to explanations from the D-N approach is to consider the following three different explanatory requests all couched in the syntactically and semantically identical expressions:

a. Why did *Ms R.* kill Mr R.?
b. Why did Ms R. *kill* Mr R.?
c. Why did Ms R. kill *Mr R.*?

The emphasis makes it clear that each question is a request for different information, and each presumably reflects differences in knowledge. Thus, the first presumes that Mr R.'s being killed needs no explanation, only why it was Ms R. instead of some other person "who done it" which needs explanation; the second question presupposes that what needs explanation is why what Ms R. did to Mr R. was a killing, and not a beating or a robbing, etc.; and the third question is a request for information that rules out other persons beside Mr R. as the victim of Ms R. The D-N model is blind to these differences in explanation, which result from these differences in emphasis. Some philosophers who rejected logical empiricism advance an account of scientific explanation that starts with pragmatics.

Following an analysis of explanation due to Bas van Fraassen, call all of these different questions expressible by the same sentence the "contrast class". Call the sentence they share in common the "topic" of the question. Now, we may associate with every question a three-membered set, whose first member is its topic, whose second is the member of the contrast class picked out by the interests of whoever requests the explanation, and whose third member is a standard for what counts as an acceptable answer to the question, which is also fixed by the interests and information of the person seeking the explanation. Call this standard on acceptable answers to our explanatory question "the relevance relation", for it determines what answers will be judged to be relevant in the context to the topic and the member of the contrast class in question. We may even identify every explanatory question with this set:

$$Q \text{ (why is it the case that } \textbf{Fab})? = \; < \textbf{Fab}, \; \{\textbf{Fab},\textbf{Fac},\textbf{Fad},\dots \}, \; \textbf{R} >$$

topic   contrast class   relevance relation

where "**Fab**" is to be read as "a bears relation **F** to **b**"; thus **Fad** means a bears relation **F** to **d**, etc. So if **F** is used to symbolize the property of " … is taller than … ", then "**Fbc**" reads, "**b** is taller than **c**". If **F** is used to symbolize the property of " … killed … ", then **Fab** means **a** killed **b**, and so on. The question **Q** above is to be understood as including whatever emphasis or other pragmatic element is necessary to make clear exactly what is being asked. For example, "Why *Ms R.* killed her husband" will be a different question from "Why Ms R. *killed* her husband", and different from "Why Ms R. killed *her husband*". All questions have (pragmatic) presuppositions ("Who let the dog escape again?" presupposes that the dog escaped and not for the first time, and that someone was responsible for allowing it). Explanatory questions are no exception. The presuppositions of **Q** include at least the following: that the topic **Fab** (the description of what is to be explained) is true; that the other possibilities (the rest of the contrast class), **Fac**, **Fad**, etc., didn't occur.

Finally, the presuppositions of **Q** include the existence of an answer to **Q**, call it **A**. **A** explains **Q** if, in light of the background knowledge of the inquirer, there is some *relationship* between **A** and the topic, **Fab**, and the rest of the contrast class (**Fac**, **Fad**, etc.) which excludes or prevents the occurrence of the rest of the contrast class, and assures the occurrence of the topic, **Fab**. In our example, we seek a true statement which, given our knowledge, bears the *relationship* to the topic and the contrast class that makes Ms R.'s killing her husband true and the members of the contrast class false. Van Fraassen calls this relationship between **A** and the topic and the contrast class "the relevance relation". We will want to know much more about this *relationship*. If our answer **A** is that Ms R. wanted to inherit Mr R.'s money, then the background knowledge will include the usual assumptions about motive, means and opportunity that are the police detective's stock in trade. If our background knowledge includes the fact that Ms R. was rich in her own right, and indeed, much richer than her husband, the relevance relation will pick out another statement, for example that Ms R. was pathologically avaricious. Of course a scientific explanation will presuppose a different "relevance relation" from that involved in the explanation of why Ms R. killed her husband. Van Fraassen tells us in effect that what makes an explanation scientific is that it employs a relevance relation fixed by the theories and experimental methods that scientists accept at the time the explanation is offered.

How does all this apparatus enable us to improve on the D-N model? Because the analysis makes explanation openly pragmatic, it has no problem with the I-S model, nor with the notion that in different contexts explaining the flagpole's height by appeal to its shadow's length will succeed. In the flagpole example, if we know about the egalitarian and patriotic desires of the city mothers of Missoula the explanation in terms of the sun's rays, the size of the shadow and the geometry of isosceles triangles will explain the height of the flagpole. Similarly, in the I-S explanation, if we don't know that Ms R. is a millionaire and/or we are acquainted with no further statistical generalizations about voting patterns, the initial I-S argument will be explanatory.

Independent of its ability to deal with the counterexamples, a pragmatic approach to explanation has its own motivation. For one thing, we might want to distinguish between a correct explanation and a good one. This is something the D-N and I-S models cannot do, but which the pragmatic account can accommodate. Some true explanations are not good ones, and many good ones are not true. An example of the first kind frequently cited in philosophy explains to a child why a square peg will not fit in a round hole by appeal to the first principles of the quantum theory of matter instead of by appeal to facts the inquirer is familiar with and can understand. An example of a good explanation if not a true one is provided by any of the well-confirmed but superseded theories which are part of the history of science. Physicists know well the defects in Newtonian mechanics. But Newtonian mechanics continues to provide explanations, and good ones at that.

But the philosopher interested in *scientific* explanations will rightly complain that no matter what its other virtues, this pragmatic account does not illuminate scientific as opposed to other kinds of (non-scientific) explanations. In effect this pragmatic analysis of explanation leaves us no clearer than we were on what makes an explanation scientific. All it tells us is that explanations are scientific if scientists offer and accept them. What we want to know are the standards for the "relevance relation" which will distinguish its explanations from the pseudo-explanations of astrology or for that matter the non-scientific explanations of history or everyday life. If we cannot say a good deal more about the relevance relation, our analysis of explanation will have little or no prescriptive bearing on how explanations ought to proceed in science, nor will it enable us to demarcate scientific from non-scientific explanations.

Summary

Our starting point for understanding scientific explanation is the deductive-nomological (D-N) or covering law model, advanced by the logical empiricists. This analysis requires that scientific explanations satisfy the requirement of giving good grounds that their *explanandum* phenomena were to be expected. If we can deduce the occurrence of the event or process to be explained from one or more laws and boundary conditions, we will have satisfied this requirement.

Thus, the requirements for scientific explanation on this view are

1 The *explanans* logically implies the *explanandum*-statement.
2 The *explanans* contains at least one general law that is required for the validity of the deduction.
3 The *explanans* must be testable.
4 The *explanans* must be true.

Several of these conditions raise serious philosophical problems.

One particularly important problem is that of exactly why laws explain. Laws are held to do so either because they report causal dependencies or alternatively because they express some sort of necessity in nature. On one account widely influential, causation just consists in law-governed sequence, so the problem becomes one of what distinguishes laws from mere accidental regularities that reflect no necessities. This apparent difference is reflected in the way laws support counterfactuals, but this difference is by itself only a symptom and not an explanation of what their necessity consists in.

Many explanations in physical science and most explanations in social science fail explicitly to satisfy this model. Exponents of D-N explanation argue that explanations can in principle do so, and they should if they are to provide real explanations. Of course many explanations approximate to the D-N model and for many purposes such "explanation sketches" are good enough.

Other philosophers reject both the D-N model and its motivation. Instead of a search for an objective standard against which to measure explanations for scientific adequacy, they focus on attempting to uncover the logic of the explanations scientists – physical, biological, social and behavioral – actually give. One reason to find this alternative strategy attractive arises when we consider the logical empiricist account of statistical explanations, the inductive-statistical, I-S, model. For whether a statistical generalization is explanatory seems to be a matter of what is known about the population, in the form of background information, by those asking for the explanation and those offering it.

But the alternative "pragmatic" approach to explanation does not successfully identify what distinguishes scientific explanations from non-scientific ones. This leads to problems we continue to explore in the next chapter.

Questions

1 Defend or criticize: "The D-N or covering law model doesn't illuminate the nature of explanation. If someone wants to know why x happened under conditions y, it's not illuminating to be told that x is the sort of thing that always happens under conditions y."
2 Supporting counterfactuals is just a symptom of the necessity of laws. In what does this necessity consist? If there is no such thing as physical or natural necessity, why do laws explain?
3 Can we directly observe causation every time we see a pair of scissors cut or a hammer pound? If we can, what philosophical problems might this solve?
4 Defend or criticize: "The D-N model represents an appropriate aspiration for scientific explanation. As such the fact that it is not attainable is no objection to its relevance for understanding science."
5 Exactly where do the pragmatic and the D-N accounts of explanation conflict? Can they both be right?

Further reading

Many important papers on explanations, causation and laws, and on all the other central topics in the philosophy of science, are reprinted in two large anthologies, R. Boyd, P. Gaspar, and J. D. Trout, *The Philosophy of Science*, and M. Curd and J. A. Cover, *Philosophy of Science: The Central Issues*. Many of the papers anthologized in these two volumes are difficult. The latter volume provides especially cogent editorial essays explaining and linking the articles.

The debate about the nature of explanation begins with classical papers written by Carl G. Hempel in the 1940s and 1950s and collected together with his later thoughts in *Aspects of Scientific Explanation*. Much of the subsequent literature of the philosophy of science can be organized around the problems Hempel raises for his own account and deals with in these essays. The final essay, from which the title of the work comes, addresses the work of other philosophers who responded to Hempel's account.

The subsequent history of debates about the nature of explanation is traced in Wesley Salmon, *Four Decades of Scientific Explanation*, originally published as a long essay in *Scientific Explanation*, vol. 13 of the Minnesota Studies in the Philosophy of Science, ed. W. Salmon and P. Kitcher. The volume from which it comes is a trove of contemporary papers on the nature of scientific explanation. Salmon has long been particularly concerned with statistical explanation, a matter treated along with other topics in his *Scientific Explanation and the Causal Structure of the World*.

Hume advanced his theory of causation in Book I of *A Treatise of Human Nature*. Its influence in the philosophy of science cannot be overstated, though few adhere to it. A latter-day empiricist account of laws is advanced by A. J. Ayer, "What is a Law of Nature", in *The Concept of a Person*. *Hume and the Problem of Causation*, by T. L. Beauchamp and the present author, expounds and defends Hume's view. J. L. Mackie, *The Cement of the Universe*, provides an exceptionally lucid introduction to the issues surrounding causation, causal reasoning, laws and counterfactuals, and defends an empiricist but non-Humean view. R. M. Tooley, *Causation: A Realist Approach*, presents a widely discussed non-empiricist approach. Miller, *Fact and Method: Explanation, Confirmation and Reality in the Natural Sciences*, defends an explicitly causal account of explanation.

Kneale, *Probability and Induction*, advances a strong and long-influential account of the natural necessity of laws. The problem of counterfactuals was first reported in N. Goodman, *Fact, Fiction and Forecast*. The most influential treatment of the nature of counterfactuals is David Lewis, *Counterfactuals*, and "Causation", in his *Philosophical Papers*, vol. 2.

van Fraassen's approach to explanation is developed in *The Scientific Image*. P. Achinstein, *The Nature of Explanation*, advances a pragmatic theory of explanation which differs from van Fraassen's.

J. Pitt, *Theories of Explanation*, reprints many important papers on explanation, including: Hempel's original paper; W. Salmon, "Statistical Explanation and Causality"; P. Railton, "A Deductive-Nomological Model of Probabilistic Explanation"; B. van Fraassen, "The Pragmatic Theory of Explanation" and P. Achinstein, "The Illocutionary Theory of Explanation".

Other important papers on explanation are mentioned in Further Reading at the end of the next chapter, also devoted to explanation.

CHAPTER 3
Scientific explanation and its discontents

Overview

Our search for the nature of scientific explanation leads us back to an examination of causes they cite and laws that connect causes to the effects they explain. An examination of causal explanation makes it clear that what we identify as the cause of an event is almost always merely one among many conditions that could bring it about, and by no means guarantees that it will happen. Moreover, most of the laws we cite include *ceteris paribus* – other-things-being-equal – clauses. This means that explanations which cite such laws, or such causes, cannot satisfy the logical positivist requirement of giving good grounds to expect their *explanandum*-event to have occurred.

The situation is perhaps graver, for *ceteris paribus* laws are difficult to subject to empirical test: we can't ever be sure that "all other things are equal". Besides such "other things being equal" laws, there are ones that report probabilities, and these come in two varieties. Some statistical generalizations, like the one examined in Chapter 2, reflect our limited knowledge and are stop-gap substitutes for strict laws. Others, like the basic laws of quantum physics, are ineliminably statistical. But such non-epistemic probabilities are difficult for empiricists about science to accept, for they do not appear to be grounded in further more fundamental processes.

Some philosophers have sought a feature of scientific explanation that is deeper than its employment of laws and commitment to reporting causal relations. They have sought the nature of explanation in the unification of disparate phenomena under deductive systems that explanations, and especially explanations of laws, often provide.

But beyond unification, people have sought even more from scientific explanations: purpose and intelligibility. The explanation of both human action and biological processes proceeds by citing their purposes or goals to explain the behavior (people work in order to earn money; the heart beats in order to circulate the blood). On the one hand, these explanations don't seem to be causal; after all, the *explanans* obtains after the *explanandum* in these cases. On the other hand, purposive explanations seem more satisfying than explanations in physics. How these "teleological" – goal-directed – explanations can be reconciled with anything like causal explanation is a problem to be addressed.

The traditional complaint that scientific explanations only tell us how something happens, and not really why, reflects the view that the complete and final explanation of things will somehow reveal the intelligibility of the universe or show that the way things are in it is the only way they could be. Historically famous attempts to show this necessity reflect a fundamentally different view of the nature of scientific knowledge from that which animates contemporary philosophy of science.

1 Inexact laws and probabilities

Answering the question of what is "the relevance relation" between questions and answers in scientific explanation brings us back to those same issues which vex the D-N model as an objective non-epistemically relativized relation between events in the world or propositions that are made true by these events. In the last decades of the twentieth century two answers to this question about the relevance relation suggested themselves. The first, due to Wesley Salmon, is a throw-back to pre-positivist approaches to scientific explanation: in a scientific explanation the relevance relation between question and answer is satisfied by those answers which reveal the causal structure that makes **A** the answer to **Q**, that treats the "because" in the statement "Fab (in contrast to the rest of the contrast class) because **Q**" as a causal relation. The second widely discussed theory of what constitutes the relevance relation for scientific explanations is due to Friedman and Kitcher. It treats the because-relation quite differently. It identifies scientific explanations as those which effect the greatest unification in our beliefs. In some respects these views are very different, and they reflect a fundamental dissensus in the philosophy of science, but in other respects they show how much the solution to problems about the nature of explanation turns on classical questions of philosophy.

The claim that what makes an explanation scientific is the fact that it is causal goes back, in some ways, to Aristotle, who distinguished four different kinds of causes. Of these, the one which science has accepted as explanatory since Newton is the notion of an "efficient cause" – the immediately prior event which gives rise to, produces, brings about what the *explanandum* describes. Physics seems to have no need for the other kinds of causes Aristotle distinguished. This is because of physics' apparent commitment to mechanism – the thesis that all physical processes can be explained by the pushes and pulls exemplified when billiard balls collide. Biology and the human sciences apparently call upon a second of the different types of causes Aristotle identified, so called "final" causes – ends, goals, purposes – for the sake of which events occur. For example, it appears to be a truth of biology that green plants use chlorophyll *in order* to catalyze the production of starch. We will return to final causes below. For the moment consider some of the problems surrounding the notion of efficient cause which we need to deal with if causation is to illuminate scientific explanation.

The first of these problems we have already alluded to: an account of the nature of causation must distinguish causal sequences from mere coincidences. If the distinction is grounded on the role of laws instantiated by causal sequences, then we need to be able to distinguish laws from accidental generalizations. It is all well and good to note that laws support counterfactuals or express physical, chemical, biological or some other sort of natural necessity, but we must not mistake these symptoms for sources of the difference between laws and accidental generalizations.

A second problem about efficient causes focuses on the actual character of causal explanations inside and outside science, one which reveals their pragmatic dimensions, their complicated relation to laws, and shows the difficulties of actually satisfying the D-N model or any account of scientific explanation like it. Suppose the lighting of a match is explained by citing its cause – that the match was struck. It is obvious that striking was not sufficient for the lighting. After all, had the match been wet, or there been a strong breeze, or no oxygen, or had the match been previously struck, or the chemical composition defective, or ... or ... or, the match would not have lit. And there is no limit to the number of these qualifications. So, if the striking was the cause, causes are at most necessary conditions for their effects. And all the other qualifications mention other necessary conditions – the presence of oxygen, the absence of dampness, the correct chemical composition, etc. But then what is the difference between a cause and a mere condition. Some philosophers have argued that the context of inquiry is what makes this distinction: in the context of an evacuated chamber used to test match heads for hardness by striking them, the cause of a match's lighting is not the striking, but the presence of oxygen (which should not be present in an evacuated chamber). Notice that this makes causal claims as pragmatic as explanatory ones are held to be. If our aim is to ground explanation on objective causal relations in the world, an account of causes that relativizes them to explanatory interests and background knowledge won't do.

If causes are but necessary conditions, then of course citing a cause will not by itself give good grounds to expect its effect. We need to be confident that the other indefinitely many positive and negative conditions that together with the cause are needed to bring about the effect also obtain. Now we can see one reason why positivists preferred to appeal to laws rather than causes as the vehicles of explanation. A law of the form "All As are Bs" or "Whenever A occurs, B occurs", or "If A, then B" fulfils the good-grounds condition since its antecedent (A) is the sufficient condition for its consequent (B). However, if laws mention sufficient conditions for their consequents, and they underwrite causal sequences, as most philosophers of science hold, then these antecedents will have to include all the necessary conditions which along with a cause bring about its effect. For instance, a law about match-strikings being followed by match-lightings will have to include clauses mentioning all the conditions, besides striking, individually necessary and jointly sufficient for match lightings. If the number of such conditions is indefinitely large, the law cannot do this, at least not if it can be expressed in a sentence of finite length that we can state. This means either that there is no law of match-striking and lighting or that if there is, its antecedent includes some sort of blanket "other things being equal" or *ceteris paribus* clause to cover all the unstated, indeed perhaps even unimagined, necessary conditions needed to make the antecedent sufficient for the lighting.

Of course there is no law about match-strikings and lightings. Rather, the laws that connect the striking to the lighting are various, large in number and mostly unknown to people who nevertheless causally explain lightings by appeal to strikings. This means that most ordinary and many scientific explanations are what we have called explanation sketches. They satisfy D-N type requirements only to the extent of presupposing that there are laws – known or unknown – which connect the boundary conditions to the *explanandum*-phenomenon. Thus explanations in the natural sciences which do not cite all the laws relevant to showing why an event occurred will be explanation-sketches, like those in history and social sciences. They are "explanation sketches", because the explainer is committed to there being some laws or other that link the boundary conditions – the cause – to the *explanandum*-event – the effect.

Moreover, if the causes laws cite are sufficient for their effects, then the scientific laws we have uncovered will also have to mention all the conditions necessary for their consequents, or else will have to contain implicit or explicit *ceteris paribus* laws. So, for example, Nancy Cartwright has argued. Thus, the inverse square law of gravitation attraction tells us that the force between two bodies varies inversely as the square of the distance between them. But we need to add a *ceteris paribus* – other things being equal – clause which will rule out the presence of electrostatic forces, or magnetic forces. There are only a small number of fundamental physical forces, so the problem of testing laws posed by *ceteris paribus* may be manageable in basic physics. But what happens when the number of conditions we need to hold constant increases greatly, as it does in biological generalizations, for example. As the number of possible interfering factors to be held constant grows, the testability of laws is reduced, making it too easy for anyone to claim to have uncovered a scientific law. This in turn threatens to trivialize causal or D-N explanation. If most of the laws we actually invoke in explanations carry implicit or explicit *ceteris paribus* clauses, then testing these laws requires establishing that other things are indeed equal. But doing so for an inexhaustible list of conditions and qualifications is obviously impossible. And this means that there are no detectable differences in kind between real laws with inexhaustible *ceteris paribus* clauses, and pseudo-laws without real nomological (i.e. law-based) force – disguised definitions, astrological principles, New Age occult theories of pyramid power or crystal magic.

For these latter "laws" too can be protected from apparent disconfirmation by the presence of *ceteris paribus* clauses. "All Virgos are happy, *ceteris paribus*" cannot be disconfirmed by an unhappy person with a mid-August birthday since we cannot establish that besides the person's unhappiness, all other things are equal. This immunity from disconfirmation, along with wishful thinking, explains the persistence of astrology.

The testability of laws is something to which we return at length in Chapter 5, but there are consequences of this problem for our understanding of how science explains. In particular, when we exchange the appeal to causes for an appeal to laws, we avoid one problem – the relativity of causal judgments – at the cost of having to deal with another – the need to deal with *ceteris paribus* clauses. The problem is all the more pressing owing to a contemporary debate about whether there are any strict laws – exceptionless general truths without *ceteris paribus* laws – anywhere in science. If the inverse square law of gravitational attraction, for example, contains a proviso excusing counterexamples resulting from the operation of Coulomb's law in the case of highly charged but very small masses, then perhaps the only laws without *ceteris paribus* clauses to be found in science are those of relativity and quantum theory.

Still another problem for those who continue to seek the nature of scientific explanation in the causal relationships such explanations report is the fact that many such relationships are increasingly reported in statistical terms. Most common are epidemiological relations like those between exposure to the sun and skin-cancers which are reported in statistical form, but which are taken to express causal relations. It is easy to say that A causes B if and only if A's presence increases the probability of B's presence, *ceteris paribus*, but now we must unpack the *ceteris paribus* clause. For we know full well that statistical correlation does not by itself warrant or reflect causal connection. But along with this problem there is a further and equally serious one. We need to understand the meaning of the concept of probability at work in causal processes. For example, it is widely accepted that smoking causes cancer because it is associated with a 40 per cent increase in the probability of lung cancer. Another sort of causal claim important in science describes how events cause changes in probabilities. For instance, an electron passing through detector A will cause the probability that another one will pass through detector B to increase by 50 per cent.

These two kinds of probabilistic causal claims are significantly different. One is meant to be a statement in part about our knowledge; the other is a claim that is supposed to hold true even when we have learned everything there is to know about electrons. Each of them makes for a different problem in our understanding of causality.

The problem with saying that smoking causes cancer, when the

probability of a smoker contracting cancer is 40 per cent and the probability of a non-smoker doing so is, say, 1 per cent, is twofold: some
smokers never contract cancer, while some lung cancer victims never
smoked. How do we reconcile these facts with the truth of the claim that
smoking causes an increase in the probability of cancer? The fact that
some lung cancer victims never smoked is not so serious a methodological
problem. After all, two effects of the same kind can have quite different
causes: a match may light as a result of being struck, or because another
already lit match was touched to it or because it was heated to the
kindling temperature of paper. The first fact, that some smokers don't
contract lung cancer, is harder to reconcile with the claim that smoking
causes cancer. One proposal philosophers have made goes like this:
smoking can be said to cause cancer if and only if, among all the different
background conditions we know about – heredity, diet, exercise, air pollu-
tion, etc. – there is no correlation between smoking and a lower than
average incidence of lung cancer, and, in one or more of these background
conditions, smoking is correlated with a higher incidence in lung cancer
rates.

Notice that this analysis relativizes causal claims to our knowledge of
background conditions. Insofar as we seek a notion of causation that
reflects relations among events, states and processes independent of us
and our theorizing about them, this analysis is unsatisfactory. But can we
just substitute "all background conditions" for "background conditions we
know about"? That would eliminate the reference to us and our knowledge. Unfortunately it also threatens to eliminate the probabilities we are
trying to understand. For "all background conditions" means the detailed
specific causally relevant circumstances of each individual who smokes.
And by the time we have refined these background conditions down to
each individual, the chance of the individual contracting cancer will turn
out to be either 0 or 1, if the underlying causal mechanism linking
smoking and specific background conditions to cancer is a deterministic
one reflecting strict laws instead of probabilities. Our probabilistic causes
will disappear. The fact that causal statements based on probabilities
reflect our available information, will be a problem for the D-N model or
any model that treats scientific explanation as a relation between statements independent of our beliefs. On the other hand, pragmatic accounts
of scientific explanation will need to be filled in, as we noted above, with
conditions on what sort of information about statistical data makes an
explanation that relies on them scientific. We cannot accept an analysis of
scientific explanation that makes anyone's answer to an explanatory
question scientifically relevant.

By contrast with probabilistic causal claims that seem to reflect limitations on our knowledge, there are the basic laws of physics, which

quantum mechanics assures us are ineradicably probabilistic. These are laws like "the half-life of U^{235} is 6.5×10^9 years", which means that for any atom of U^{235} the probability that it will have decayed into an atom of lead after 6.5×10^9 years is .5. Laws like these do not merely substitute for our ignorance, nor will they be replaced through refinement to strict non-probabilistic ones. Quantum mechanics tells us that the fundamental laws operating at the basement level of phenomena are just brute statements of probability, which no further scientific discoveries will enable anyone to reduce or eliminate in favor of deterministic strict laws. The law about the half-life of uranium attributes to uranium atoms a tendency, a **disposition**, a propensity to decay at a certain probabilistic rate. But the probabilities these laws report present us with still another difficulty for causation. The causal probabilities of quantum mechanics are "tendencies", "dispositions", "capacities", "propensities" or powers of some subatomic arrangements to give rise to others.

These probabilistic powers are troublesome to some scientists and many philosophers. This is because dispositions can really only be understood by explaining them in terms of further more basic non-dispositions. To see this, consider a non-probabilistic disposition, say, fragility.

A glass is fragile if and only if, were it to be struck with sufficient force it would shatter. But, note, this is a counterfactual statement, and it will be accepted only if there is a law which supports it, a law which either reports a causal relationship between glass being fragile and shattering when struck. And this law about fragile objects obtains owing to a causal relationship between the molecular structure of glass and its shattering when struck. All (normal) glasses are fragile, but many glasses never shatter. Their fragility consists in their having the molecular structure reported in the law which supports the counterfactual. In general, attributing a disposition, or capacity or power to something is tantamount to hypothesizing a causal relationship between some of that things' non-dispositional, structural properties and its behavior. Being fragile is having a certain structure, a structure which the object has all the time, even when it is not being struck or shattering. Here is another example: a piece of metal's being magnetic is a matter of attracting iron filings, and its being a magnet consists in the arrangement of atoms that make it up in a lattice, and the orientation of the electrons in these atoms. This arrangement is present in a magnet, for example, even when it is not exerting a magnetic force on anything nearby.

Applying this result to the **probabilistic propensities** quantum mechanics reports is problematical. Since these probabilities are propensities or dispositions, and are the most fundamental "basement"-level properties physics reports, there cannot be a more basic level of structural properties to causally ground these probabilities. They are therefore "free-floating" powers of microphysical systems, which the systems probabilistically manifest, but which when not manifested, exist without any actual further causal basis. Compare fragility or magnetism: can these potentials be present in a glass or a piece of iron without some actual property to underlie them – such as molecular composition, or orientation of outer-shell electrons in a lattice? Without such a "base" we cannot understand probabilistic propensities as dispositions, powers or capacities with causal foundations. We cannot establish their existence as distinct from their effects – the frequencies with which the quantum effects they bring about occur. There is nothing to show for them independent of our need to somehow ground probabilistic regularities at the basement level of physics. These pure probabilistic dispositions will be very different from the rest of the dispositional causes that science cites to explain effects. Unlike fragility or magnetism or any other disposition science studies, quantum probabilistic propensities are beyond the reach of empirical detection (direct or indirect) independent of the particular effects they have. They have all the metaphysical mysteriousness of the concept of causal or nomological necessity.

These are some of the problems which must be addressed by those who seek to ground scientific explanation on the notion of causation. It may now be easier to see why many philosophers have hoped to find an analysis of the nature of explanation in science which avoided having to face intractable questions about the nature of causality. One such alternative approach to explanation goes back at least to an insight of Albert Einstein's, according to which scientific theorizing should "aim at complete coordination with the totality of sense-experience" and "the greatest possible sparsity of their logically independent elements (basic concepts and axioms)". The demand for "sparsity" is translated into a search for unification.

In terms of specifying the relevance relation, between question and answer, that makes an explanation scientific, a scientific explanation will be one that effects unifications, reduces the stock of beliefs we need to have in order to effect explanations. The two key ideas are: first, scientific explanations should reflect the derivation of the more specific from the more general, so that the stock of basic beliefs we need is as small as possible. Second, which stock of basic beliefs we embrace is constrained by the need to systematize experiences. Unification is the aim of scientific explanation because, on this view, human understanding of the world increases as the number of independent *explananda* we need decreases. So, in the explanation of general phenomena, what makes an explanation scientific is that phenomena are shown to be special cases of one or more even more general processes; in the explanation of particular events, states and conditions, what makes for scientific explanation is that the *explanantia* on the one hand apply widely to other *explananda*, and that the *explanantia* themselves be unified with other beliefs by being shown to be special cases of other more general *explanantia*. According to Philip Kitcher, one of the chief exponents of this view of scientific explanation, the demand for unification makes logical deduction an especially important feature of scientific explanations, for this is what unification consists in. We shall return to the role of deduction in explanation when we examine the nature of theories in Chapter 4. Kitcher also requires that the propositions that effect unification pass stringent empirical tests. These two conditions on unification show that this alternative still shares important similarities with the D-N model of explanation. But it purports to go deeper that Hempel's general criterion of adequacy (that the *explanans* give good grounds to expect the *explanandum*) to some underlying feature of scientific explanation.

Unification does seem to contribute to understanding. But let us ask why. What makes a more compendious set of beliefs about nature better than a less compendious set, assuming that both account for the evidence – data, observations, experiences, etc. – equally well? One answer might be that the universe is simple, that the underlying causal processes that give rise to all phenomena are small in number. In that case, the search for unifications will reduce to the search for causes, and the criterion of explanatory relevance unification sets out will be a variant of the causal criterion we have set out above. If causation is, as empiricists have long held, a matter of laws of increasing generality, and if the universe reflects a hierarchy of more basic and more derived causal sequences, then explanations which effect unification will also uncover the causal structure of the world.

Now, suppose that the universe's causal structure is permanently hidden from us, because it is too complex or too small or because causal forces operate too fast for us to measure or are too strong for us to discern. But suppose further that we nevertheless can effect belief-unifications which enable us to systematize our experiences, to predict and control up to levels of accuracy good enough for all our practical purposes. In that case, for all its practical pay-off, unification will not enhance understanding of the way the world works, or will do so only up to certain limits.

Exponents of unification may have a more philosophically tendentious argument for distinguishing unification from causation and preferring it. They may hold, along with other philosophers of science, that beyond observations the causal structure of the world is unknowable and so drops out as an epistemically relevant criterion on the adequacy of explanations. More radically, they may hold (as Kitcher does) that causation consists in explanation, or that causation, like explanation, is also unification-dependent. So, unification is all scientific understanding can aim at. We will return to these issues in our discussion of the nature of theories in Chapter 4.

2 Causation and teleology

Whether scientific explanation is causal, unificatory, nomological, statistical, deductive, inductive, or any combination of them, a question may still remain about how and whether scientific explanations really answer our explanatory questions, really convey the sort of understanding that really satisfies inquiry. One very long-standing perspective suggests that scientific explanation is limited, and in the end unsatisfying, because it does not go deep enough to the bottom of things. Sometimes this perspective expresses itself in the thesis that scientific explanations only reveal how things come about, but not why they happen. Thus, for example, it will be held that all a D-N model tells us about an *explanandum*-event is that it happened because such an event always happens under certain conditions and these conditions obtained. When we want to know why something has happened, we already know that it has, and we may even know that events like it always happen under the conditions in which it happened. We want some deeper insight than how it came to happen.

When this sort of dissatisfaction with scientific explanation is expressed, what sort of explanation is sought? These deeper explanatory demands seek an account of things which show them and nature in general to be "intelligible", to make sense, to add up to something, instead of just revealing a pattern of one damn thing after another. Traditionally, there seem to be two sorts of explanations that aim at satisfying this need for deeper understanding than push–pull, "efficient"-cause explanations that physics and chemistry can provide.

Sometimes, the demand is for an explanation which will show that what happened had to happen in a very strong sense, that its occurrence was necessary, not just physically necessary, in light of what just the laws of nature just happen to be, but necessary as a matter of rational intelligibility or logic. Such an explanation would reveal why things couldn't have turned out any other way, because, for example, the laws of nature are not contingently true about the world, but necessarily true – that there is only one possible way the world can be. On this view, gravity cannot, as a matter of logical necessity, vary as the cube of the distance between objects as opposed to the square, copper must as a matter of logic alone be a solid at room temperature, the speed of light couldn't be 100 miles an hour greater than it is, etc. This is a conception of science that goes back to the eighteenth-century rationalist philosophers Leibniz and Kant, who set themselves the task of showing that the most fundamental scientific theories of their day were not just true, but necessarily true, and thus provided the most complete form of understanding possible.

There is a second sort of explanatory strategy that seeks to respond to the sense that causal explanations are unsatisfying. It goes back much further than the eighteenth-century philosophers, back past Aristotle, though he identified the kind of explanatory strategy in question. This is the notion of "final cause" explanations, which are common in biology, the social and behavioral sciences, history and everyday life. In these contexts, explanations proceed by identifying the end, goal, purpose for the sake of which something happens. Thus, green plants have chlorophyll *in order to* produce starch, Caesar crossed the Rubicon *in order to* signal his contempt for the Roman Senate, the Central Bank raised interest rates *in order to* curb inflation. In each of these cases, the explanation proceeds by identifying an effect "aimed at" by the *explanandum*-event, state or process, which explains it. These explanations are called "teleological", from the Greek, "telos", meaning end, goal, purpose. There is something extremely natural and satisfying about this form of explanation. Because it seems to satisfy our untutored explanatory interests, it serves as a paradigm for explanations. To the extent non-purposive explanations fail to provide the same degree of explanatory satisfaction, they are stigmatized as incomplete or otherwise inadequate. They don't give us the kind of "why" that final cause, purposive explanations do.

Both the attractions of an explanation that shows what happened had to happen as a matter of logical necessity which allows for no alternative, and the appeal of **teleological explanations**, are based on very controversial philosophical theses – claims which most philosophers have repudiated. If these two sorts of explanation are based on questionable assumptions, then it will turn out that despite the feeling that it isn't enough, "efficient" causal explanation will be the best science or any other intellectual endeavor can offer.

Teleological explanations seem to explain causes in terms of their effects. For example, the heart beating – the cause – is explained by its circulating the blood – the effect. Since the time of Newton, such explanations have been suspect to philosophers, because, in the words of the seventeenth-century philosopher Spinoza, they "reverse the order of nature", making the later event – the effect – explain the earlier event – the cause. If future events do not yet exist, then they cannot be responsible for bringing about earlier events. Physics does not allow for causal forces (or anything else for that matter) to travel backwards in time. Moreover, sometimes a goal which explains its cause is never attained: producing starch explains the presence of chlorophyll even when the absence of CO_2 prevents the green plant from using chlorophyll to produce starch. Thus, physical theory itself rules out the possibility of teleological explanation in physics – to the extent that teleology requires the future to determine the past.

There thus seem to be three possibilities: If physics does not allow "final causes", then either there are none, or biological and other apparently teleological processes are irreducibly different from physical processes, or, despite their appearance, when we really understand how they work, teleological processes are not really different from efficient causal processes, they just look different. On this third alternative, once we understand how teleological processes work, we will discover that they are just complicated causal processes.

The first two alternatives are philosophically controversial: it's hard to deny that things in nature have purposes, and drawing distinctions between the methods of physics and biology is likely to be disadvantageous to biology. The third alternative is worth exploring first. Can explanations that appeal to purposes really turn out to be garden variety causal explanations of the same kind as physics employs?

It is widely claimed that teleological explanations of human actions common in everyday life are unproblematic because they are really just garden variety causal explanations, in which the causes are desires and beliefs. These explanations only look teleological because the desires and beliefs are about future states or events or conditions, and they are identified in terms of these future states. Thus, my buying a ticket on the Eurostar on Monday is explained by the desire to go to Paris next Friday. But this desire took hold last Sunday. There is no future causation here, though there is a description of the prior cause – the desire felt on Sunday – in terms of its "content" – the future effect of my going to Paris on Friday. If these explanations are causal, then presumably there is a law or laws that link desires and beliefs on the one hand as causes to actions on the other hand as effects. Many explanations and theories in social science presuppose that there is such a law, one expressed in the theory of rational choice: "agents choose that action among attainable ones which will secure their strongest desire, other things equal". Whether the theory of rational choice, as developed by economists for example, is a *bona fide* general law is a separate question from the role which it is accorded in providing causal explanations in the social sciences, history and ordinary life. In these areas its explanatory adequacy is unchallenged.

Desire/belief–action explanations go back further in our culture than any recorded history. They are the explanations we employ to explain and justify our own actions. And when we put ourselves in the shoes of those whose actions we seek to understand, desire/belief–rational-choice *explanantia* provide a kind of "intelligibility" for their *explananda* lacking in the natural sciences. Uncovering the desires and beliefs that animate a person's action gives them meaning. It is this or a similar notion of meaning that is missing from the explanations natural science provides.

So rational-choice explanations are in the end causal and not really teleological: if the desires and beliefs they cite are the causes of the actions they explain, then it cannot be teleology that makes for the complete explanatory satisfaction they seem to provide. Rather, it will be the sort of "intelligibility" or "meaning" which desire/belief explanations provide for actions, and which is missing from physical science. If the meaning or intelligibility that rational-choice explanations provide turns on the operation of a causal law connecting beliefs and desires with actions, then there will in the end be no difference in kind between the explanation of human action and explanation in physics. And where desires and beliefs do not come into play, in physics and chemistry and biology and the rest of natural sciences, the demand for a more satisfying form of explanation that reveals the meaning of things will be unfounded.

It is partly for this reason that there is a long-standing debate in the philosophy of psychology and the philosophy of social science about exactly how desires and beliefs explain actions, and whether they do so causally or not. If desire/belief rational-choice explanation is after all non-causal, then, first of all, meanings cannot be captured causally; second, human action can't be treated scientifically; and, finally, the search for meanings beyond human affairs, if there are any, must transcend natural science.

It is often at this point that religious and other non-scientific attempts to explain natural phenomena enter. By insisting that the demand for meaning or explanatory intelligibility is always in order even for physical events and processes, they undermine the claims of science to provide complete explanations, or indeed any satisfying explanations of things. If natural processes are not the result of human desires and beliefs, then the only fully satisfactory explanation of them is to be given by superhuman, divine will, by a God. This God's "desires" and "beliefs" – his or her will, omniscience and omnipotence – bring about and give meaning to the events for which natural science can only trace out the immediate prior causes.

In biology, at least until the middle of the nineteenth century, the hypothesis that crucial facts about organisms were to be explained in this particularly satisfying way was a reasonable one. Before Darwin's theory of natural selection, arguably the likeliest explanation for the adaptatedness and complexity of biological organization was provided by appeal to God's design – one rendered biological organization intelligible by giving the meaning of the parts biologists identified in terms of their roles in a plan. However, with the advent of Darwin's theory of evolution as the result of heritable variation blind to adaptive needs, and natural selection that filters out the less well adapted, the hypothesis that adaptation and complexity is the result of design becomes significantly less likely. What the theory of natural selection shows is that the appearance of design could be the result of purely causal processes in which no one's purposes, goals, ends, intentions, will, etc., play any part. Thus, green plants bear chlorophyll because at some point or other through blind variation their predecessors were endowed with chlorophyll, the endowment was inheritable, chlorophyll happens to catalyze the production of starch, producing starch keeps plants alive. And that's why latter-day plants have chlorophyll. The "in-order-to" of our original explanation gets cashed in for an etiology in which the filter of natural selection culls those plants that lacked chlorophyll or its chemical precursors, and selects for those which have it, or which mutate from the precursor closer and closer to chlorophyll as it exists in present-day green plants. As for where the first precursor molecules came from, on which nature selects and selects and selects until chlorophyll emerges, that first precursor is the result of purely non-foresighted chemical processes to be explained by chemistry without recourse to its adaptative significance for plants.

If we consider the evidence for Darwinian theory sufficient, we must conclude not just that the appearance of design could have been produced without the reality of design, but that there is no deity whose plan gives rise to the adaptation and complexity of biological systems, there is no meaning, nor any real intelligibility to be found in the universe. There may remain room in the scientist's **ontology** for a deist's conception of God as the first cause, but no room for cosmic meanings endowed by God's interventions in the course of nature.

Thus, the demand that something more be provided than causal explanations, something which will render nature intelligible or give its processes meaning, showing why things happen in the sense of giving their teleology, is just unwarranted in the light of modern science. The demand for meaning rests on a factual mistake about the nature of the universe. We know it for a mistake because, as the eighteenth-century physicist Laplace said, in answer to the King of France's question about the place of God in his system, "Your majesty, I have no need of this hypothesis." If we can explain the how of everything – their efficient causes – and there is no sign that things fit into someone's plan, there is no scope for the sort of dissatisfaction with causal explanations that people who seek "the meaning of it all" sometimes express.

The philosopher who argues in this way, of course, is taking sides on a substantive scientific question: whether or not we need to hypothesize further forces, things and processes to explain nature than those so far countenanced by science. Since science is both incomplete and fallible, we cannot rule out that further evidence, or indeed previously misinterpreted evidence, will lead us to conclude that such further non-physical factors are required and that they may show things to have meaning or intelligibility beyond anything we have hitherto supposed. Philosophers who read the claims of science differently, or place credence in non-scientific considerations, will differ from those who reject as unwarranted the dissatisfaction which regards causal explanations as in principle inadequate to provide complete understanding.

3 From intelligibility to necessity

We are left with the first of our two sources of dissatisfaction with causal explanation: the idea that it does not provide for intelligibility in a different sense from that of design and meaning, a sense in which intelligibility is the demonstration that the course of nature is necessary: that there is a sense in which there is no other way things could have turned out. Many philosophers and others have held the view that scientific explanation should uncover underlying mechanisms responsible for the course of nature which are necessarily true. Two important eighteenth-century philosophers, Leibniz and Kant, argued that science does in fact reveal such necessities. As such, science's explanations are complete, leave nothing unexplained, allow for no alternative account, and therefore bear the highest degree of certainty, as befits science. Leibniz sought to show that once physical knowledge was complete we would see that each law fitted together with the rest of scientific theory so tightly that a difference in one would unravel the whole structure of scientific theory. The inverse square law of gravitational attraction could not have been an inverse cube law without some other law having been different, and differences in that law would make for further differences in other laws until we discovered that one or more of the changes leads to logical contradiction and incoherence. Hence, the package of completed science – all the laws – will be logically necessary, and there is a kind of logical inevitability to the way in which the course of nature they govern is played out. Leibniz did not argue for this view by showing exactly how changes in our best scientific theories actually do ramify into incoherence. He could not do so because scientific knowledge was in his time too incomplete even to try. It is still too incomplete to show any such incoherence. Moreover, even if we acquired a package of scientific laws that work together to explain all phenomena, we will need some assurance that this is the only package of scientific laws that will do so. The logical consistency of all our scientific laws – indeed, their arrangement in a deductive order that unifies them in an axiomatic system – is by itself insufficient to rule out the existence of another such axiomatic system, with different axioms and theorems, which effect the same systematization of phenomena. This is the problem of "**underdetermination**", to which we shall return in the next chapter. Meanwhile, what Leibniz's conception requires is an argument that the axioms of this logically coherent package which explains everything are themselves necessary truths. This is something Kant aimed to supply.

Kant argued that at least a good deal of the foundations of Newtonian mechanics are necessary truths, and set out to show how this is possible. His theory holds that the nature of space and time, the existence of a cause for every physical event – causal determinism – and, for example, the Newtonian principle of the conservation of matter, are necessary because they reflect the only way in which cognitive agents like us can organize our experiences. As such these principles can be known "*a priori*" – independently of experience, observation, experiment – through the mind's reflection on its own powers – pure reason. Unlike Leibniz, Kant recognized that scientific laws are not logical truths. By contrast with the laws of logic, and with statements true by definition, like "All bachelors are unmarried males", the denial of a scientific law is not self-contradictory. Employing a distinction Kant introduced and which has remained important in philosophy since the eighteenth century, true propositions, such as scientific laws, whose denials are not self-contradictory, are "**synthetic truths**", by contrast with "**analytic truths**". Kant defined these truths as ones whose subject "contains the predicate", for example, "all bachelors are adult unmarried males". "Contains" is obviously a metaphor, but the idea is that analytic truths are statements true by definition or the consequences of definitions. Analytic truths, as definitions or their deductive consequences, are without content, make no claims about the world, and merely indicate our stipulations and conventions about how we will use certain noises and inscriptions. For example, "density equals the quotient of mass and volume" makes no claim about the world. It does not imply that there is anything that has mass, volume or density. The definition cannot explain any fact about the world, except perhaps facts about how we use certain noises and inscriptions. If "having a certain density" could explain why something has a certain mass-to-volume ratio, it would be a case of "self-explanation" – an event, state or condition explaining its own occurrence. For having a certain density just is having a certain mass-to-volume ratio. If nothing can explain itself, analytic truths have no explanatory power. A synthetic truth, by contrast, has content, makes claims about more than one distinct thing or property in the world, and thus can actually explain why things are the way they are. The laws of nature are thus synthetic truths.

Kant accepted that Newton's laws were universal truths and that they were necessary truths as well. Since he held that universality and necessity are the marks of *a priori* truths, Kant set out to explain how it is possible for the fundamental laws of nature to be "synthetic *a priori* truths". That is, how they can make explanatory claims about the world that we can know without recourse to observation, experiment, the collection of data or other sensory experiences. If Kant's program of establishing the synthetic *a priori* character of, say, physics, had succeeded, then its explanations would have a special force beyond simply telling us that what happens here and now does so because elsewhere and elsewhen events of the same kind happen in circumstances of the kind that obtain here and now. According to Kant, the special force such explanations bear consists in these being the only explanations our minds can by their very nature understand, and their truth is also assured to us by the nature of human thought itself. Pretty clearly, explanations of this character will be particularly satisfying, not to say exhaustive and exclusive of alternatives.

Kant believed that unless he could establish the synthetic *a priori* truth of at least physics, it would be open to skeptical challenge by those who deny that humans can discover natural laws, and those who hold that the laws we can uncover do not reveal the essential nature of things. In particular, Kant was concerned to refute an argument he identified as David Hume's: If the laws of nature are not knowable *a priori*, then they can only be known on the basis of our experience. Experience, however, can provide only a finite amount of evidence for a law. Since laws make claims that, if true, are true everywhere and always, it follows that their claims outrun any amount of evidence we could provide for them. Consequently, scientific laws are at best uncertain hypotheses, and the claims of physics will be for ever open to skeptical doubt. Moreover, Kant feared that speculative metaphysics would inevitably seek to fill this skeptical vacuum.

Kant was correct in holding that the laws of nature are synthetic. However, for the philosophy of science the most significant problem facing Kant's account of Newtonian theory as synthetic truths known *a priori* is that the theory isn't true at all, and so cannot be known to be true *a priori*. What is more, its falsity was established as the result of experiment and observation. And since these experiments and observations underwrite theories, notably Einstein's theories of relativity, and quantum mechanics incompatible with Newton's theory, neither Newton's laws nor their successors could in fact be known *a priori*. Philosophers of science concluded that the only statements we can know *a priori* will be those which lack content, i.e. definitions and the logical consequences of definitions which do not constrain the world at all, and so have no explanatory relevance to what actually happens. Since experience, observation, experiment, etc. can never establish the necessity of any proposition, scientific claims with explanatory relevance to the way the world actually is cannot be necessary truths. Two important consequences follow from this conclusion. First, the search for an alternative to causal explanation that reveals the necessity or intelligibility of the way things are is based on a misunderstanding: necessary truths have no explanatory power. Second, if a proposition has any explanatory power, if it is a statement with content, in Kant's term "synthetic", and not "analytic", then it can only be justified by observation, experiment, the collection of data.

This conclusion, however, leaves us confronted with Hume's problem: since the empirical evidence for any general law will always be incomplete, we can never be certain of the truth of any of our scientific laws. But Hume raises an even more serious "problem of induction". He begins by noting that if we could be certain that the future will be like the past – that is, the uniformity of nature – then our past experiences would support scientific laws. But unless pure reason alone can vouchsafe the future uniformity of nature, the only assurance we can have that the future will be uniform with the past is our past experiences of its uniformity hitherto. Hume pointed out that pure reason cannot do this. There is no ground to suppose that in the future nature will be uniform with the past. After all, the denial of the uniformity of nature is no contradiction (imagine tomorrow fire being cold and ice being warm). But our past experience will justify our beliefs about the future only if we already have an independent right to believe that in the future nature will be similar to – uniform with – the past. If the evidential relevance of past experience to shape future expectations requires that nature be uniform, past experience of nature's uniformity cannot underwrite this requirement. It would be tantamount to asking to borrow money on the verbal promise to repay, and then when the reliability of one's promise is challenged, to enhance its credibility by promising that one will keep the promise. If the reliability of one's promises is at issue, using a promise to certify their reliability won't do. If relying on the future uniformity of nature to assure inferences from the past to the future is at issue, it won't do to say, such inferences from past to future have always been reliable up till now – for this is to infer from past reliability to reliability at the next opportunity. This is Hume's "problem of induction". It is treated at greater length in Chapter 5.

Hume's argument is widely taken to claim at least that science is inevitably fallible, and, more radically, that scientific knowledge cannot be justified by experience at all. If Hume is right, the conclusions of scientific investigation can never acquire the sort of necessity required by Kant, Leibniz and others who have craved certainty or necessity. But this fallibility will be unavoidable in any body of scientific laws which have explanatory content, which make claims about the way the world works.

Hume's problem of induction is a problem for philosophers. No scientist can afford to suspend empirical inquiry until it is solved. In fact, the problem is best treated as a reflection of the central role which empirical testing plays in securing scientific knowledge. A statement which can figure in a scientific explanation must be testable by experience. This requirement, that the claims of science must be testable, is both the most widely accepted conclusion and the source of the most intractable problems in the philosophy of science.

Summary

Most scientists will agree about which explanations are good ones and which are not. In the previous chapter we saw that the trouble comes when we make serious attempts to express precisely the standards they implicitly employ and to find general features common and distinctive of good scientific explanations. We may all agree that such explanations must involve laws. But to begin with, the exact degree of involvement of laws – explicit participation, implicit backing or something in between – is open to dispute. And then there is the problem of telling scientific laws from other sorts of non-explanatory generalizations and sorting out why the former explain when the latter do not. This problem turns out to involve the philosopher's mystery about whether there are real necessities in nature. If there are no such necessities, it becomes hard to see what explanatory content laws have over what we have called merely accidental generalizations. If laws do have the kind of necessity which makes them explanatory, it is not a property they openly manifest for all to recognize. Indeed, there is the fundamental problem of telling how close our best guesses are to the laws of nature. Unless we can tell, we will have no basis on which to say whether any of our putative explanations do more than merely reduce temporary onslaughts of curiosity. Attempts to avoid many of these problems, by turning our attention from laws to, say, causes as the bearers of explanatory power in science, will be not only unavailing but rather ironical. For it was to laws that the logical empiricists appealed in order to avoid traditional problems about causation. For example, they hoped to trade the problem of what causal necessitation consists in for an account of the difference between general laws and accidental generalizations. But these two problems turn out to be the same.

In this chapter we recognized that causes are usually at most necessary prior conditions, not sufficient conditions which guarantee their effects, and most if not all laws reflect this fact by their *ceteris paribus* – other-things-equal – clauses.

Probabilistic laws seem to come in two types. There are the ones that summarize the state of our partial knowledge of phenomena instead of identifying their causes; and then there are the probabilistic laws of quantum physics with their unexplainable probabilistic propensities – that is, dispositions to behave in ways that can be given numerical probabilities without these values being based on any further facts about the things that have these dispositions. If both sorts of laws explain, then scientific explanation may not be a single homogeneous process.

Scientific explanation has traditionally met with dissatisfaction by those who demand that such explanation show the purpose, design or meaning of natural processes, and not just the processes which show how they came to happen. This demand for final cause or teleological explanation goes back to Aristotle. Contemporary accounts of teleological explanation exploits Darwin's discovery of how blind variation and natural selection can give rise to purpose or perhaps only the appearance of purpose. Whether Darwin's theory excludes purpose from nature or merely naturalizes it, the theory helps us see that teleological explanation is only a complex and disguised form of causal explanation.

Relatedly, there is a tradition which goes back at least to the seventeenth-century philosopher Leibniz and the eighteenth-century philosopher Kant, of arguing that scientific explanation must ultimately show that science's description of reality is not just true, but necessarily, logically true. That the way the world is, is the only way it could be. We have good reason to think that any attempt to establish such a conclusion is bound to fail. Indeed, were it to succeed, we would be hard pressed to explain much of the fallible and self-correcting character of scientific knowledge.

One question we have not yet settled is the matter of general strategy in the philosophy of science: Do we treat science in the way Plato would have, as a set of interrelated propositions about the world that obtain independent of us and that we set out to discover, or we treat science as a human creation, an invention, not a discovery, so that its fundamental character is as much a reflection of our interests and styles of thought as it is a mirror of nature. Each of these age-long perspectives animates a different philosophy of science. They will recur forcefully in the next chapter on the nature of theories, and force us to make a choice among aims for science between which no fully satisfying compromise is possible.

Questions

1 If, as some philosophers argue, all laws have *ceteris paribus* clauses, what implications are there for limits to explanation, and to prediction?

2 Defend or criticize: "The fact that scientific explanation cannot provide for the intelligibility or necessity of things, is a good reason to seek it elsewhere."
3 Does The Darwinian theory of natural selection show that there is no such thing as purpose in nature or does it show that there are purposes and they are perfectly natural causal processes?
4 Why is it difficult for empiricists to accept quantum mechanical probabilities as fundamental unexplainable facts about the world?
5 How different is the D-N model from the view that scientific explanation is a matter of unifying disparate phenomena?

Further reading

Aristotle advanced his theory of four causes in the *Physics*. The problem of *ceteris paribus* clauses is treated insightfully in one of Hempel's last papers, "Provisos", in A. Grunbaum and W. Salmon, *The Limitations of Deductivism*. Nancy Cartwright, *How the Laws of Physics Lie*, is the *locus classicus* for arguments that all laws bear *ceteris paribus* clauses.

J. L. Mackie, *Truth, Probability and Paradox*, includes two exceptionally clear essays from an empiricist perspective on the meaning of probability statements and on the problem of dispositions. W. Salmon, *The Foundations of Scientific Inference*, provides an excellent account of Hume's problem of induction, as well as the prospects for alternative interpretations of probability to solve it. K. Popper defends a probabilistic propensity interpretation of quantum mechanics in *Objective Knowledge*.

Kitcher's defense of explanation as unification is to be found in W. Salmon and P. Kitcher, *Scientific Explanation*, as well as a paper anthologized in Pitt, *Theories of Explanation*. This anthology also contains a paper developing the same view independently by M. Friedman.

The way in which Darwinian theory can be used to assimilate purpose and teleology to causation is most influentially explained in L. Wright, *Teleological Explanation*. An anthology, C. Allen, M. Bekoff and G. Lauder, *Nature's Purposes*, brings together almost all of the important papers on this central topic in the philosophy of biology. The nature of intentional explanation in the social sciences is treated in A. Rosenberg, *Philosophy of Social Science*.

Much of Leibniz's work remains untranslated, and what is available is very difficult. Perhaps most valuable to read in the present connection is *New Essays on Human Understanding*. I. Kant, *The Critique of Pure Reason*, defends the claim that the most fundamental scientific theories are synthetic truths known *a priori*. Hume's problem of induction is to be found in his *Inquiry Concerning Human Understanding*, which also develops Hume's account of causation and his defense of epistemological empiricism.

CHAPTER 4
The structure and metaphysics of scientific theories

Overview

Overview

How often have you heard someone's opinion written off with the statement, "that's just a theory". Somehow in ordinary English the term "theory" has come to mean a piece of rank speculation or at most a hypothesis that is doubtful, or for which there is as yet little evidence. This usage is oddly at variance with the meaning of the term as scientists use it. Among scientists, so far from suggesting tentativeness or uncertainty, the term is often used to describe an established subdiscipline in which there are widely accepted laws, methods, applications and foundations. Thus, economists talk of "game theory" and physicists of "quantum theory", biologists use the term "evolutionary theory" almost synonymously with evolutionary biology, and "learning theory" among psychologists comports many different hypotheses about a variety of very well established phenomena. Besides its use to name a whole area of inquiry, in science "theory" also means a body of explanatory hypotheses for which there is strong empirical support.

But how exactly a theory provides such explanatory systematization of disparate phenomena is a question we need to answer. Philosophers of science long held that theories explain because, like Euclid's geometry, they are deductively organized systems. It should be no surprise that an exponent of the D-N model of explanation should be attracted by this view. After all, on the D-N model, explanation is deduction, and theories are more fundamental explanations of general processes. But unlike deductive systems in mathematics, scientific theories are sets of hypotheses, which are tested by logically deriving observable consequences from them. If these consequences are observed, in experiment or other data collection, then the hypotheses which the observations test are tentatively accepted. This view of the relation between scientific theorizing and scientific testing is known as "**hypothetico-deductivism**". It is closely associated with the treatment of theories as deductive systems, as we shall see.

The axiomatic conception of theories naturally gives rise to a view of progress in science as the development of new theories that treat older ones as special cases, or first approximations, which the newer theories correct and explain. This conception of narrower theories being "reduced" to broader or more fundamental ones, by deduction, provides an attractive application of the axiomatic approach to explaining the nature of scientific progress.

Once we recognize the controlling role of observation and experiment in scientific theorizing, the reliance of science on concepts and statements that it cannot directly serve to test by observation becomes a grave problem. Science cannot do without concepts like "nucleus","gene",

"molecule", "atom", "electron", "quark" or "quasar". And we acknowl-
edge that there are the best of reasons to believe that such things exist.
But when scientists try to articulate their reasons for doing so, difficulties
emerge – difficulties borne of science's commitment to the sovereign role
of experience in choosing among theories.

These difficulties divide scientists and philosophers into at least two
camps about the metaphysics of science – realism and **antirealism** – and
they lead some to give up the view of science as the search for unifying
theories. Instead, these scientists and philosophers often give pride of
place in science to the models we construct as substitutes for a complete
understanding that science may not be able to attain. We need to identify
what is in dispute between those who see science as a sequence of useful
models and those who view it as a search for true theories.

1 How do theories work?

What is distinctive about a theory in this latter sense is that it goes beyond the explanations of particular phenomena to explain these explanations. When particular phenomena are explained by an empirical generalization, a theory will go on to explain why the generalization obtains, and to explain its exceptions – the conditions under which it fails to obtain. When a number of generalizations are uncovered about the phenomena in a domain of inquiry, a theory may emerge which enables us to understand the diversity of generalizations as all reflecting the operation of a single or small number of processes. Theories, in short, unify, and they do so almost always by going beyond, beneath and behind the phenomena empirical regularities report, to find underlying processes that account for the phenomena we observe. This is probably the source of the notion that what makes an explanation scientific is the unifications it effects. For theories are our most powerful explainers, and they operate by bringing diverse phenomena under a small number of fundamental assumptions.

For the philosophy of science the first question about theories is how do they effect their unifications. How exactly do the parts of a theory work together to explain a diversity of different phenomena? One answer has been traditional in science and philosophy since the time of Euclid. Indeed, it is modeled on Euclid's own presentation of geometry. Like almost all mathematicians and scientists before the twentieth century, Euclid held geometry to be the science of space and his "elements" to constitute a theory about the relations among points, lines and surfaces in space.

Euclid's theory is an axiomatic system. That is, it consists in a small set of postulates or axioms – propositions not proved in the axiom system but assumed to be true within the system – and a large number of theorems derived from the axioms by deduction in accordance with rules of logic. Besides the axioms and theorems there are definitions of terms, such as straight line – nowadays usually defined as the shortest distance between two points – and circle – the locus of points equidistant from a given point. The definitions of course are composed from terms not defined in the axiomatic system, like point and distance. If every term in the theory were defined, the number of definitions would be endless, so some terms will have to be undefined or "primitive" terms.

It is critical to bear in mind that a statement which is an axiom that is assumed to be true in one axiomatic system, may well be a theorem derived from other assumptions in another axiom system, or it may be justified independently of any other axiom system whatever. Indeed, one set of logically related statements can be organized in more than one axiom system, and the same statement might be an axiom in one system and a theorem in another. Which axiomatic system one chooses in a case like this cannot be decided by considerations of logic. In the case of Euclid's five axioms, the choice reflects the desire to adopt the simplest statements which would enable us conveniently to derive certain particularly important further statements as theorems. Euclid's axioms have always been accepted as so evidently true that it was safe to develop geometry from them. But, strictly speaking, to call a statement an axiom is not to commit oneself to its truth, but simply to identify its role in a deductive system.

It is clear how Euclid's five axioms work together to systematize an indefinitely large number of different general truths as logically derived theorems. Thus, if we measure the internal angles of a triangle with a protractor, the result will always approach 180 degrees. The explanation of why follows pretty directly from the axioms of Euclid: they enable us to prove that the interior angles of a triangle equal exactly 180 degrees. First we prove that when a line is drawn between two parallel lines, the alternate angles of intersection are equal.

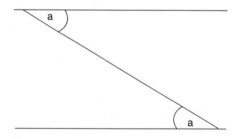

Add to this the theorem that a straight line is equal to a 180 degree angle, and we can demonstrate that therefore the internal angles of a triangle can be added up to equal a straight line.

It is easiest to give the proof by a diagram:

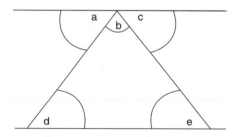

Notice that angle **a** = angle **d**, and angle **c** = angle **e**, while angle **b** is equal to itself. Since the upper line is a straight line, it is a 180 degree angle, and the sum of angles **a**, **b** and **c** equals 180 degrees too. But then the sum of angle **d**, angle **b** and angle **e** must also equal 180 degrees. Thus we demonstrate that the interior angles of a triangle equal 180 degrees.

Each proof in geometry illustrates a different way in which Euclid's axioms work together to enable us to derive a theorem – one we can independently confirm by construction or measurement of shapes and solids, and which also explains why these solids and shapes have the features we can measure or construct. But because there are indefinitely many such theorems, there are indefinitely many ways that these axioms work together, and we can give no general account of what working together comes to, beyond saying that in Euclid's theory, and in scientific theories generally, the axioms work together to explain general phenomena in logically deductive arguments. The trouble with this claim is that it goes almost nowhere towards illuminating the notion of components of theories "working together". Consider the following "theory" composed of two axioms "working together" and the theorems deduced from them:

The ideal-gas law:

$$PV = rT$$

where P = pressure, T = temperature and V = volume, and r is the universal gas constant.

The quantity theory of money:

$$MV = PT$$

where M is the quantity of money in an economy, V = the velocity of money – the number of times it changes hands – P is the average price of goods and T is the total volume of trade.

From the conjunction of these two laws, either one of them follows logically, by the simple principle that if "A and B" then "A". And so do other generalizations. For example, from $PV = rT$ and some definitions it follows that when the pressure on the outside of a balloon is constant, an increase in temperature increases its volume. From the quantity theory of money it follows that, other things being equal, increasing the amount of money in circulation results in inflation. Yet clearly, our theory as a whole does not in any way explain the processes that follow logically from either of its axioms alone.

In a theory, the parts must work together to explain. But working together cannot be captured by the notion of logical derivation alone. Yet saying exactly what it is about the components of a theory that make it one theory, instead of several joined together, is the beginning of another long-standing philosophical challenge. For the philosopher of science it is not enough to say simply that a theory is a body of laws that work together to explain. "Works together" is too vague. More important, philosophers of science seek to clarify what it is exactly about a theory that enables it to do the scientific work it does – explaining a large number of empirical regularities, and their exceptions, and enabling us to predict outcomes to greater degrees of precision than the individual laws which the theory subsumes.

One natural suggestion emerges from the conclusion of Chapters 2 and 3. The fundamental, underived general laws of a theory work together by revealing the causal structure of underlying processes that give rise to the laws which the theory systematizes and explains. So, what's wrong with the theory composed of the ideal-gas law and the quantity theory of money is that there is no single underlying structure common to both the behavior of gases and money for there to be a theory about. How do we know this? Presumably because we know enough about gases and money to know that they have nothing directly to do with one another. But even concepts like underlying causal structure or mechanism may not provide the degree of illumination we seek. Chapter 2 uncovered some serious reasons why philosophers are reluctant to place too much weight on the notion of causation. What is worse, the notion of an underlying mechanism may seem disturbing, given the empiricist argument that there is nothing to causation beyond regular sequence – no glue, no mechanism, no secret powers or necessities in nature to link events together in ways that make the course of things inevitable or intelligible. With these reminders about difficulties ahead and behind, we must nevertheless explore the notion that a theory is a body of laws that work together to explain phenomena by attributing an underlying causal structure or mechanism to the phenomena. We must do so because so many theories manifestly work like this.

Perhaps the philosopher's favorite example of a theory is the so-called kinetic theory of gases, according to which (a) gases are made of molecules moving on straight-line paths, (b) molecules – like observable objects – are governed by Newton's laws of motion, except that (c) molecules are perfectly elastic, take up no space and, except when they collide, exert no gravitational or other forces on one another. Given these assumptions it is relatively easy to explain the ideal-gas law, **PV = rT**. The trick is to connect the underlying structure – the behavior of molecules *like* billiard balls – with the measurements we make of the gas's temperature, pressure and volume. One of the important discoveries of nineteenth-century thermodynamics consists in effecting this connection: When a gas is in equilibrium,

The absolute temperature of gas = $\frac{1}{2}mv^2$

where **m** is the mass of an individual molecule and **v** is the average velocity of the ensemble of molecules that the gas in the container is composed of. $\frac{1}{2}mv^2$ is the standard expression for kinetic energy in Newtonian mechanics. Here it is attributed to unobservable molecules which are treated *as though* they were inelastic spheres – billiard balls – that collide. By recognizing that heat and pressure are the macroscopic reflections of molecular motion, physicists were able to explain the gas laws – laws which had been known since the time of Boyle and Charles in the seventeenth century. If we set temperature equal to the mean kinetic energy of the molecules of the gas, and pressure equal to the momentum transferred per cm^2 per second to the sizes of the container by the molecules as they bounce off it, we can derive the ideal-gas law (and other laws it subsumes – Boyle's law, Charles's law, Guy Lussac's law) from Newton's laws applied to molecules. We can also derive Graham's law, according to which different gases diffuse out of a container at rates which depend on the ratio of the masses of their molecules, and Dalton's law that the pressure one gas exerts on the walls of a container is unaffected by the pressure any other gas in the container exerts on it. We can even explain Brownian movement – the phenomenon of dust motes in the air remaining in motion above the ground never dropping towards the ground under the force of gravity: they are being pushed in random paths by collision with gas molecules that compose the air. There is in principle no end to the regularities about different types, amounts and mixtures of particular gases we can derive from and thereby explain by the kinetic theory of gases.

Let's generalize a bit from this case. The kinetic theory of gases consists in Newton's laws of motion plus the law that gases are composed of perfectly elastic point-masses (molecules) which obey Newton's laws, plus the law that the temperature of a gas (in degrees Kelvin) is equal to the mean kinetic energy of these point masses, plus some other laws like this one about pressure and volume of the gas.

The kinetic theory thus explains observable phenomena – the data we collect when we measure changes in temperature and pressure of a gas, holding volume constant; or pressure and volume, holding temperature constant, etc. The theory does so by making a set of claims about invisible, unobservable, undetectable components of the gas and their equally unobservable properties. It tells us that these components and their properties are governed by laws that we have independently confirmed as applying to observable things like cannon balls, inclined planes, pendula and, of course, billiard balls. The kinetic theory thus provides an example of one way the components of a theory work together to explain observations and experiments.

The kinetic theory of gases can illustrate several further components of an approach to the nature of theories that emerged naturally out of the deductive-nomological or covering law approach to explanation which we elaborated in Chapter 2. This approach is usually described nowadays as the axiomatic or syntactic account of scientific theories. It is associated with a view of the way theories are tested known as "hypothetico-deductivism", according to which scientists theorize – frame hypotheses – but do not test them directly, because like most theories in science they are typically about processes we cannot directly observe. Rather, the scientist deduces testable consequences from these hypotheses. If the tests are borne out by observation, the hypotheses are (indirectly) confirmed. Thus, the axiomatic or **syntactic approach to theories** is sometimes called the "hypothetico-deductive" or H-D account of theories.

The axiomatic approach begins with the notion that theories are, as we have suggested, axiomatic systems, in which the explanation of empirical generalizations proceeds by derivation or logical deduction from axioms – laws not derived but assumed in the axiomatic system. Because the axioms – the underived laws fundamental to the theory – usually describe an unobservable underlying mechanism – like our point-mass billiard-ball-like gas molecules – they cannot be directly tested by any observation or experiment. These underived axioms are to be treated as hypotheses *indirectly* confirmed by the empirical laws derivable from them, which can be directly tested by experiment and observation. It is from these two ideas, that the foundations of a theory are *hypotheses* supported by the consequences *deduced* from them, that the name hypothetico-deductive model derives.

One theory's underived axioms are another theory's explained theorems, of course. Every theory leaves something unexplained – those processes which it invokes to do the explaining. But these processes unexplained in one theory will presumably be explained in another. For example, the balanced equations of chemical stoichiometry (for example $2H_2 + O_2 \rightarrow 2H_2O$) are explained by assumptions the chemist makes about electron-sharing between hydrogen and oxygen atoms. But these laws, underived in chemistry, are the derived, explained generalizations of atomic theory. And atomic theory's assumptions about the behavior of electrons which result in the chemical bond are themselves derived in quantum theory from more fundamental generalizations about the components of microparticles. No one suggests that scientists actually present theories as axiomatic systems, still less that they explicitly seek the derivations of less fundamental laws from more fundamental ones. It is important to remember that like the covering law model, the axiomatic account of theories is a "rational reconstruction" of scientific practice designed to reveal its underlying logic. Nevertheless, it claims to have found vindication both in the long-term history of science, and in important theoretical breakthroughs of recent science.

Consider the accomplishments of Watson and Crick, the molecular biologists who discovered how the chemical structure of the chromosome – the chains of DNA molecules of which it is composed – carry hereditary information about traits from generation to generation. Watson and Crick's theory about the molecular structure of the gene enables geneticists to explain the fundamental underived laws of Mendelian genetics – laws about how hereditary traits, like eye color, are distributed from generation to generation. How did this happen? In principle the situation is little different from the derivation of the ideal-gas law, $PV = rT$, from the kinetic theory of gases: given the identification of the gene with a certain amount of DNA, the laws governing the segregation and assortment of genes from generation to generation should be logically derivable from a set of laws governing the behavior of DNA molecules. One reason this should be so is of course that the Mendelian gene is nothing but a strand of DNA – that is what Watson and Crick discovered. So, if there are laws about Mendelian genes it stands to reason that they obtain in virtue of the operation of laws about DNA molecules. And if this is so, then how more clearly to show that one set of laws obtains in virtue of another set of laws than to logically derive the former from the latter. Indeed, if we could not at least in principle do so, there would seem to be good reason to think the Mendelian laws to be independent of and autonomous from the "lower level laws". Since the lower level ones explain the higher level laws, these cannot be independent of the lower level ones. Logical derivation formalizes this explanatory relation.

This process whereby more basic or fundamental theories explain less general ones, improve on them, deal with their exceptions, and unify our scientific knowledge, has seemed to many philosophers of science to characterize the history of science since the time of Newton. For some millennia before Newton it was widely held by scientists and non-scientists alike that the motion of heavenly bodies, the planets and stars was governed by fixed laws and the motion of things on and near the earth governed either by no laws or by another set of laws quite different from those governing heavenly motion. This belief reflected an even more fundamental conviction that the realm of the heavens was perfect, unchanging, incorruptible and entirely different in material composition from the realm of the earth. Here on earth things were thought to happen in irregular ways that show few patterns, things break down and disorder continually threatens to take over, things grow and die. In short, the earth was supposed to be a far less perfect world than the heavens. The connections between this scientific world-view and that of the dominant religions before the scientific revolution are obvious.

The accomplishments of Kepler, Galileo and Newton in the sixteenth and seventeenth centuries completely overthrew this world-view and replaced it with a metaphysics that reflected their theoretical accomplishment. And at the core of this accomplishment was the discovery of the laws of celestial and terrestrial motion by Kepler and Galileo respectively and the logical derivation of both of their sets of laws from a single more fundamental set of laws by Newton.

Employing data gathered by the Danish sixteenth-century astronomer Tycho Brahe, Kepler showed that we could predict the position of the planets in the night sky by assuming that they travel around the sun on ellipses and that their velocity is a specific function of their distance from the sun. Since we are "aboard" one of these planets its actual motion and that of the other planets around the sun is hidden from us, but the confirmation of predictions about the apparent position of the planets in the night sky provides indirect confirmation for Kepler's hypothesis about elliptical orbits.

Galileo's experiments, dropping cannonballs, according to legend, from the leaning tower of Pisa, rolling them down inclined planes, timing the period of pendula as their lengths are changed, all contributed to his discovery of the laws of motion of objects in the immediate vicinity of the earth: projectiles always follow the paths of parabolas, pendula periods depend on the length of the wire and never the weight of the bob, freefalling bodies of any mass have constant acceleration.

It was Newton's achievement to show that Kepler's laws of planetary motion and Galileo's laws of terrestrial motion both can be derived from a single set of four laws: the law that bodies on which no forces act are at rest (i.e. have zero acceleration, and if they have non-zero velocity, they move in straight lines), that force is equal to the product of mass and acceleration, the conservation law of equal and opposite action and reaction, and the inverse square law of gravitational attraction. In showing that Kepler's laws and Galileo's were but special cases of more general laws true everywhere and always, Newton not only explained why their laws obtained, he also undercut the basic metaphysical conviction that the realm of the heavens was somehow different from that of the earth.

Along with Galileo's telescopic discovery of the craters and other imperfections of the moon, Newton's revolution had a profound intellectual influence far beyond the formal derivation which he provided to unify physical theory. Moreover, the power of Newton's unification was further sustained in the ensuing two hundred years as more and more phenomena came to be explained (or explained in more precise quantitative detail) by it: eclipses; the period of Halley's Comet; the shape of the earth – a slightly squashed sphere; the tides; the precession of the equinoxes; buoyancy and aerodynamics; parts of thermodynamics; were unified and shown to be "the same underlying process" through the derivation of laws describing these phenomena from Newton's four fundamental laws.

Philosophers of science refer to this derivation of the laws of one theory from the laws of another as "intertheoretical reduction" or simply "**reduction**". Reduction requires that the laws of the reduced theory be derived from that of the reducing theory. If explanation is a form of derivation, then the reduction of one theory to another explains the reduced theory; in effect it shows that the axioms of the less basic theory are theorems of the more basic one.

So the scientific revolution of the seventeenth century appears to consist in the discovery and reduction of Galileo's and Kepler's laws to Newton's, and the progress of physics from the sixteenth century onwards is the history of less general theories being successively reduced to more general theories, until the twentieth century, when suddenly theories even more general than Newton's are framed, which in turn reduce Newtonian mechanics by derivation: the special and general theories of relativity and quantum mechanics. Newton's laws are deducible from the laws of these theories by making some idealizing assumptions, in particular that the speed of light is infinite or at least that all other attainable velocities are much, much slower than the speed of light, and the idealizing assumption that energy comes in continuous amounts and not in discrete but very small units or "quanta".

According to one traditional view in the philosophy of science, the reduction of theories to more fundamental ones reflects the fact that science is successively enlarging its range and depth of explanation as more and more initially isolated theories are shown to be special cases, derived from a smaller and smaller number of more fundamental theories. Scientific change is scientific progress, and progress comes in large measure through reduction. In fact, reduction is also viewed as the characteristic relation among disciplines once they attain the status of sciences. Thus, in principle, chemistry should be reducible to physics, and biology should be reducible to chemistry via molecular biology. Similarly, we should seek a psychological science composed of laws themselves reducible to the laws of biology. Of course, the social sciences have yet to or never will, uncover laws reducible to those of natural science, via reduction to psychological laws. Therefore, these disciplines lack an important feature common to scientific theories – linkage via reduction to the most fundamental and predictively powerful of the sciences, physics.

We can now understand some of the attractiveness of axiomatization as an account of how a theory explains by uncovering more general under- lying mechanisms that systematize and explain less general ones. If the universe reflects the neat picture of layers of causal laws, each of which rests on a layer of laws below it that logically imply these laws, and if the universe is composed of a small number of basic kinds of things that behave in a uniform way and out of which everything else is composed, then there should be a uniquely correct description of nature which will take axiomatic form because reality is a matter of the complex being built up out of the simple in accordance with general laws. The commitment to axiomatization as giving the structure of theory and the relations among theories is tantamount to a metaphysical claim about the nature of reality: at bottom it is simple in composition and operation, and all the complexity and diversity of more complicated and more composite things is the result of the simplicity at the bottom of things.

Of course, this picture must be substantially complicated. To begin with, the notion that the laws of one theory may be directly derivable from those of another is too simple. Scientific progress involves the correction and improvement of a theory's predictions and explanations by its successors. If the successor theory merely "contained" the original reduced theory as a logical consequence, it would incorporate the errors of its predecessor. For example, Galileo's law of terrestrial motion implies that the acceleration of bodies falling towards the earth remains constant, while Newton's laws recognize that accelerations must increase owing to the gravitational force between the earth and bodies approaching it. For predictive purposes we can neglect these slight increases in acceleration, but we must correct Galileo's terrestrial mechanics, adding gravitational force, if it is to follow from Newton's laws. Similarly, Mendel's laws of genetics should not follow directly from laws in contemporary molecular genetics, for we know that Mendel's laws are wrong. Phenomena like genetic linkage and gene-crossover falsify these laws. What we want of any reduction of Mendel's laws to more fundamental laws of molecular genetics is an explanation of where Mendel's laws go wrong as well as where they work. This suggests that reduction usually involves deriving a "corrected" version of the theory to be reduced from the more funda- mental reducing theory.

But the requirement that the reduced theory must sometimes be "corrected" creates problems for the axiomatic view of theory change. Sometimes, one theory supersedes another not by reducing it, but by replacing it. Indeed, replacement seems characteristic of a discipline's becoming a "real" science. For example, Lavoisier's oxygen theory did not reduce the older phlogiston theory of combustion. It replaced the "ontology" – the kinds of things phlogiston theory was about: calx, phlogiston, dephlogisticated air, etc. – and its alleged laws, by providing a completely different kind of thing, oxygen, which could not be linked up to phlogiston in ways that would enable this latter concept to survive in Lavoisier's theory of combustion. Accordingly, scientists say that there never was any such thing as phlogiston. By contrast, when a theory is reduced to a broader or more fundamental one, the "ontology" of the reduced theory – the kinds of things it makes claims about, is preserved. The reason is that reduction is a matter of deduction of the law of the reduced theory from those of the reducing theory, and such derivation is possible only when the terms of the two theories are connected. You can't derive the laws of Mendelian genetics from those of molecular genetics unless the Mendelian gene can be defined in terms of nucleic acids. For it is assemblages of DNA which molecular genetics are about and Mendelian genes which Mendel's laws are about: a law about all **C**s being **F**s will only follow from a law about all **A**s being **B**s if every **A** is identical to a **C** and every **B** is identical to an **F**. Indeed, a large measure of the achievement of reduction is the formulation of these identities. For example, the reduction of the thermodynamics of gases to statistical mechanics turns on the identity we noted above:

Absolute temperature at equilibrium = $\frac{1}{2}mv^2$

Whether we treat this identity as a definition or a general law relating temperature and kinetic energy, its formulation was the crucial breakthrough that enabled physicists to reduce the behavior of gases to the behavior of the molecules which compose them.

It seems a characteristic feature of reduction that it unifies observable phenomena or at least unifies the generalizations that report them to more and more fundamental, more and more accurate regularities which are more and more observationally inaccessible. Having begun with cannonballs and planets, physics succeeds finally in explaining everything in terms of undetectable microparticles and their properties. So, it seems to make explanatorily basic what is epistemically most problematical – hardest to acquire knowledge of. While the official epistemology of science is empiricism – the thesis that our knowledge is justified only by experience, that is, experiment and observation, its explanatory function is fulfilled by just those sorts of things that creatures like us can have no direct experience of. Indeed, the microparticles of modern high energy physics are things no creature like us could have acquaintance with. And this fact raises the most vexing questions about the nature of scientific theories.

2 The problem of theoretical terms and the things they name

Scientific explanations are supposed to be testable, they have "empirical content", their component laws describe the way things are in the world and have implications for our experience. But almost from the outset science has explained by appeal to a realm of untestable entities, processes, things, events and properties. As far back as Newton, physicists and philosophers have been uncomfortable about the fact that such things seem both necessary and unknowable. Unknowable, because unobservable; necessary because without appeal to them theory cannot effect the broad unification of observations that the most powerful explanations consist in. Gravity is a good example of the problem.

Newtonian mechanics makes sense out of a vast range of physical processes by showing how they are the result of contact between bodies with mass. We can explain the behavior of a wind-up clock, for example, by tracing a causal chain of cogs, wheels, weights, hour and minute hands, chimes and twittering birds in which the pushes and pulls observations detect are quantified and systematized into exchanges of momentum and conservation of energy between things in contact with one another. And this mechanical explanation will itself presumably give way to an even more basic explanation in terms of the mechanical properties of the component parts of the cogs and wheels, and in turn the mechanical properties of their parts until at last we have explained the behavior of our clock in terms of the behavior of the molecules and atoms that compose it. This at any rate is the explanatory expectation of the reductionist.

By contrast, Newtonian gravity is not a "contact" force. It is a force that is transmitted across all distances at infinite speed apparently without any energy being expended. It moves continually through total vacuums, in which there is nothing to carry it from point to point. Unlike anything else it is a force against which nothing can shield us. And yet it is a force itself completely undetectable except through its effects as we carry masses from areas of greater gravitational force (like the earth) to areas of lesser gravitational force (like the moon). All in all, gravity is a theoretical entity so different from anything else we encounter in our observations, that these observations do not help us much to understand what it could be. And it is a thing so different from other causal variables that one might be pardoned for doubting its existence, or at least being uncomfortable about invoking it to explain anything. One would not be surprised by a centuries-long search for some "mechanical" explanation of how gravity works or even better some less mysterious substitute for it.

Most of Newton's contemporaries felt this discomfort with the notion of gravity, and some followers of Descartes tried to dispense with it altogether. But neither they nor later physicists were prepared to dispense with the notion. For dispensing with gravity means giving up the inverse square law of gravitational attraction,

$$F = \frac{g\,m_1\,m_2}{d^2}$$

and no one is prepared to do this. Gravity thus seems an "occult" force, whose operation is no less mysterious than those which non-scientific explanations like astrological horoscopes invoke to allay our curiosity. And the same may be said of other such unobservable notions. Thus, the molecules which compose a gas are supposed to have the properties of little billiard balls, for it is their billiard-ball-like behavior which explains the ideal-gas law. But if gas molecules are small masses, then surely they are colored, for nothing can be a mass unless it takes up space, and nothing can take up space unless it has some color. But individual molecules have no color. So, in what sense could they be small masses? The obvious answer is that unobservable things aren't just small versions of observable things; they have their own distinct properties – charge, quantized angular momentum, magnetic moments, etc. But how do we know this if our knowledge is justified only by what we have sensory experience of? And, as noted above, by what right can we claim that theories invoking these theoretical entities and properties provide real explanations when we can have no experience of them whatever. Why should a theory about electrons or genes we cannot see, touch, smell, taste or feel be any better at explanation than astrology, New Age mystery-mongering, superstition or fairy-tales?

We can express our problem of justification as one about the meaning of words and the learnability of language. Consider the terms we employ to describe our experiences: the names for observable properties of things – their colors, shapes, textures, smells, tastes, sounds. These terms we understand because they name our experiences. Then there are the terms that describe objects that have these properties – tables and chairs, clouds and clocks, lakes and trees, dogs and cats, etc. We can agree on the meaning of these terms too. Furthermore, it is tempting to suppose that all the rest of our language is somehow built up out of the names for sensory properties and the labels for everyday objects. For otherwise, how could we have ever learned language? Unless some words are defined not by appeal to other words, but by the fact that they label things we can directly experience, we could never learn any language. Without such extra-linguistically defined terms we could not break into a never-ending circle or regress of definitions of one word by reference to other words, and those words defined by reference to still other words, and so on. We would already have to know a language in order to learn it.

Furthermore, language is an infinite disposition: we can produce and can understand any of an indefinite number of different sentences. Yet we can do so on the basis of a finite brain that has learned to speak in a finite amount of time; it is hard to see how we managed this feat unless either language is somehow innate or there is some basic vocabulary from which all the rest of language is built up. Now the hypothesis that language (as opposed to a language-learning device) is innate is one empiricists and most scientists have never taken very seriously. We were not born knowing any language; otherwise it would be hard to see how it is that any human child can with equal facility learn any human language, from birth. That leaves the hypothesis that we learned a finite stock of basic words of one language which together with composition rules enables us to build up the capacity to produce and understand any of an infinite number of sentences of that language. What else could this finite stock be but the basic vocabulary we learned as infants. And this vocabulary is of course the names of sensory experiences – hot, cold, sweet, red, smooth, soft, etc. – along with words like mum, and dad.

But if this is the basis of language, then every word with a meaning in our language must ultimately have a definition in terms of words that name sensory properties and everyday objects. And this requirement should include the theoretical terms of modern science. If these words have meaning, then they must somehow be definable by appeal to the fundamental vocabulary of experience. This argument goes back to eighteenth-century British empiricist philosophers like Berkeley and Hume. These philosophers were troubled by the "secret powers" like "gravity" and unobservable things like "corpuscles" invoked in seventeenth-century physics. Their disquiet about these theoretical entities has had a continuing impact on the philosophy of science right up to the end of the twentieth century and beyond it.

The twentieth-century followers of the British empiricists labeled themselves positivists and logical empiricists (we encountered them as proponents of the D-N model of scientific explanation in Chapter 2). The logical empiricists inferred from arguments about the learnability of language like this one that the theoretical vocabulary of science had ultimately to be "cashed in" for claims about what we can observe, on pain of otherwise being simply empty, meaningless noises and inscriptions. These philosophers went further and argued that much of what in the nine-teenth and twentieth centuries passed for scientific theorizing could be shown to be meaningless nonsense, just because its theoretical terms were not translatable into the terms of ordinary sensory experience. Thus, Marx's dialectical materialism, and Freud's psychodynamic theory were stigmatized as pseudo-science, because their explanatory concepts – surplus value, the oedipal complex, etc. – could not be given empirical meaning. Similarly, a whole host of biological theories which postulated "vital forces" were denied explanatory power by these philosophers because they invoked entities, processes, forces which could not be defined by appeal to observations. But it was not just pseudo-science which these empiricist philosophers attacked. As we have seen, even such indispen-sable terms as "gravity" were subject to criticism for lack of "empirical content". Some logical positivists, and the later nineteenth-century physi-cists who influenced them, also denied the meaningfulness of concepts such as "molecule" and "atom". For such empiricists a concept, term or word had empirical content only if it named some thing or property we could have sensory awareness of.

Of course, empiricists held there would be no problem invoking theo-retical entities if the terms we used to name them could be defined by way of observable things and their properties. For in that case not only would we be able to understand the meaning of theoretical terms, but we could always substitute statements about observables for ones about unobserv-ables if any doubt were raised. For example, consider the theoretical concept of density. Every type of material has a specific density, and we can explain why some bodies float in water and some do not by appeal to their densities. But the density of a thing is equal to its mass divided by its volume. If we can measure a thing's mass, on a scale, in a pan-balance, or some other way, and we can measure its dimensions with a meter stick, we can calculate its density: That means we can "explicitly define" density in terms of mass and volume. In effect "density" is just an "abbreviation" for the quotient of mass and volume. Whatever we say about density we could say in terms of mass and volume. It may be more of a mouthful, but the empirical content of a claim about the mass of an object divided by its volume would be the same as the empirical content of any claim about its density. So, if we could explicitly define theoretical terms by way of observable ones, there would be no more trouble understanding what they mean than there is understanding what observable terms mean. There would be no chance of a theory introducing some pseudo-scientific term in a non-scientific theory that provides merely apparent explanatory power. Most important of all, we would know exactly under what observational conditions the things named by our observationally defined terms were present or not, and were having the effects which theory tells us they do.

Unfortunately, hardly any of the terms that name unobservable properties, processes, things, states or events are explicitly definable in terms of observable properties. Indeed, the explanatory power of theories hinges on the fact that their theoretical terms are not just abbreviations for observational ones. Otherwise, theoretical statements would simply abbreviate observational statements. And if they did that, theoretical statements could summarize, but not explain, observational ones. Since density is by definition identical to mass divided by volume, we could not appeal to their differing densities to explain why two objects of equal volume are of unequal mass; we would simply be repeating the fact that their ratios of mass to volume are unequal. More important, unlike "density", few theoretical terms can even be equated with some finite set of observable traits or properties of things. For example, temperature-changes cannot be defined as equal to changes in the length of a column of mercury in an enclosed tube, because temperature also varies with changes in the length of a column of water in an enclosed tube, and changes in the resistance of an ohm-meter, or the shape of a bi-metallic bar, or changes in the color of a heated object, etc. What is more, temperature changes occur even when there are no observable changes in the length of mercury or water in a tube. You cannot employ a conventional water or mercury thermometer to measure temperature changes smaller than about .1 degree centigrade, nor to measure temperatures that exceed the melting-point of glass or fall below the freezing-point of mercury or water or alcohol or whatever substance is employed. In fact there are some things whose temperatures change in ways that no thermometer we could currently design would record. So, some physical properties or changes in them do not seem to be observationally detectable. The situation for more theoretical properties than temperature is even murkier. If an "acid" is defined as a "proton-donor" and no observations we can make give "empirical content" to the concept of a "proton-donor" because we cannot touch, taste, see, feel, hear or smell a proton, then "acid" is a term with no meaning. On the other hand we may define acid as "whatever turns red litmus paper blue", but then we won't be able to explain why some liquids do this and others don't.

Could we provide empirical meaning for the theoretical claims of science by linking complete theoretical statements with entire observable statements, instead of just individual theoretical terms with particular observable terms? Alas, no. The statement that the mean kinetic energy of the molecules in a particular gas-container increase as pressure increases is not equivalent to any particular statement about what we can observe when we measure its temperature, owing to the fact that there are many different ways of measuring temperature observationally, and that using any one of them involves substantial further theoretical assumptions about the operation of thermometers, most especially the theoretical statement that absolute temperature at equilibrium equals mean kinetic energy.

The question we face cuts right to the heart of the problem about the nature of science. After all, the "official epistemology" of science is some form of empiricism, the epistemology according to which all knowledge is justified by experience: otherwise the central role of experiment, observation and the collection of data in science would be hard to explain and justify. In the long run, scientific theorizing is controlled by experience: progress in science is ultimately a matter of new hypotheses which are more strongly confirmed than old ones as the results of empirical tests come in. Science does not accept as knowledge what cannot be somehow subject to the test of experience. But at the same time, the obligation of science to explain our experience requires that it go beyond and beneath that experience in the things, properties, processes and events it appeals to in providing these explanations. How to reconcile the demands of empiricism and explanation is the hardest problem for the philosophy of science, indeed for philosophy as a whole. For if we cannot reconcile explanation and empiricism, it is pretty clear that it is empiricism that must be given up. No one is going to give up science just because its methods are incompatible with a philosophical theory. We may have to give up empiricism for rationalism – the epistemology according to which at least some knowledge we have is justified without empirical test. But if some scientific knowledge is derived not from experiment and observation but, say, rational reflection alone, then who is to say that alternative world-views, myths or revealed religion, which claim to compete with science to explain reality, will not also claim to be justified in the same way.

The logical empiricist insists that we can reconcile empiricism and explanation by a more sophisticated understanding of how theoretical terms can have empirical content even though they are not abbreviations for terms that describe observations. Consider the concepts of positive and negative charge. Electrons have negative charge and protons positive ones. Now, suppose someone asks what the electron lacks that the proton has in virtue of which the former is said to have a negative charge and the latter is said to have a positive charge. The answer of course is "nothing". The terms "positive" and "negative" used in this context don't represent the presence and absence of some thing. We could just as well have called the charge on the electron positive and the charge on the proton negative. These two terms function in the theory to help us describe differences between protons and electrons as they *manifest themselves in experiments* we undertake with things we can observe. Electrons are attracted to the positive pole of a set of electrically charged plates and protons to the negative one. We can "see" the effects of this behavior in the visible tracks in cloud chambers or the gas bubbling up through the water in a chemical electrolysis set-up. The terms "positive" and "negative" make systematic contributions to the theory in which they figure, contributions that are cashed in by the observational generalizations which the theory of atomic structure organizes and explains. The "empirical meaning" of the term "negative" is given by the systematic contribution which the term makes to the generalizations about what we can observe that follow from the assumptions of the theory about electrons being negatively charged. Remove the term from the theory, and the theory's power to imply many of these generalizations will be destroyed, the observations it can systematize and explain will be reduced. Whatever is the extent of the reduction in explanatory power, that is what constitutes the empirical meaning of the term "negative".

We can identify the empirical content of the term "electron" or "gene" or "charge" or any other term in our corpus of theories which names an unobservable thing or property in the same way. Each must make some contribution to the predictive and explanatory power of the theory in which it figures. To identify this contribution, simply delete the term from the theory and trace out the effects of the deletion on the theory's power. In effect, "charge" turns out to be defined "implicitly" as whatever it is that has the observable effects we lose when we delete the term "charge" from atomic theory, and similarly for any other theoretical term in any theory.

This in effect is the way in which the axiomatic approach to theories dealt with the problem of theoretical terms. Logical empiricists sought to reconcile the explanatory power of the theoretical machinery of science with the constraints observation places on science by requiring that legitimate theoretical terms be linked to observations through "**partial interpretation**" – interpretation is a matter of giving these terms empirical content, which may be quite different from the words scientists use to introduce them. Interpretation is partial because observations will not exhaust the empirical content of these terms, else they lose their explanatory power.

Another example may help. Consider the term "mass". Newton introduced this term with the definition "quantity of matter", but this definition is unhelpful because matter turns out to be as "theoretical" a notion as mass. Indeed, one is inclined to explain what matter is by appeal to the notion of mass, matter being anything that has any amount of mass. Mass is not explicitly defined in Newton's theory at all. It is an undefined term. Instead of being defined in the theory, other concepts are defined by appeal to the concept of mass; for example, momentum, which is defined as the product of mass and velocity. But mass's empirical content is given by the laws in which it figures and their role in systematizing observations. Thus, mass is partially interpreted as that property of objects in virtue of which they make the arms of pan balances drop when placed upon them. We can predict that a mass coming into contact vertically with a pan balance will result in the balance arm moving because motion is the result of force, and force is the product of mass and acceleration, and moving a mass onto a pan balance causes the pan to have non-zero acceleration.

We should of course distinguish the "empirical meaning" of a term from its dictionary definition or semantic meaning. "Mass" is certainly a term with an English dictionary definition, even though its empirical meaning is quite different and it is an undefined term in Newtonian mechanics.

So, the partial interpretation of mass is provided by the means we use to measure it. But these means do not define it. For one thing, the ways we measure mass by measuring its effects, like the motion of pan balance arms, that mass causally explains. For another, there are many different ways of measuring mass by its effects, including some ways we may not yet have discovered. If such as-yet-undiscovered ways of measuring mass exist, then our interpretation of "mass" cannot be complete; it must be partial. And again, a complete interpretation in terms of observations would turn "mass" into an abbreviation for some set of observational terms, and would deprive it of its explanatory power.

The logical empiricists advanced this claim that the unobservable terms of science need to be linked by meaning to observational terms, so that the really explanatory apparatus of science could be distinguished from pseudo-explanations which attempt to trade on the honorific title of scientific theory. Ironically, they were also the first to recognize that this requirement could not be expressed with the precision their own standards of philosophical analysis required. The first half of the twentieth century's philosophy of science was devoted to framing what came to be known as a "principle of **verification**" – a litmus test which could be unambiguously applied to distinguish the legitimate theoretical terms of science from the illegitimate ones. Strong versions of the principle required complete translation of theoretical terms into observable ones. As we have seen, this requirement cannot be met by most of the terms invoked in scientific explanations; moreover, we wouldn't want theoretical terms to satisfy this requirement because if they did so, they would lose their explanatory power with respect to observations.

The problem was that weaker versions of the principle of verification preserve the dross with the gold; they fail to exclude as meaningless terms everyone recognizes as pseudo-scientific, and will not discriminate between real science and New Age psychobabble, astrology or for that matter religious revelation. It is too easy to satisfy the requirement of partial interpretation. Take any pseudo-scientific term one likes; provided one adds a general statement containing it to an already well-established theory, the term will pass muster as meaningful. For example, consider the hypothesis that at equilibrium a gas *is bewitched* if its absolute temperature equals the mean kinetic energy of its molecules. Added to the kinetic theory of gases, this hypothesis makes the property of "being bewitched" into a partially interpreted theoretical term. And if one responds that the term "is bewitched" and the added "law" make no contribution to the theory, because they can be excised without reducing its predictive power, the reply will be made that the same can be said for plainly legitimate theoretical terms, especially when they are first introduced. What after all did the concept of "gene" add to our understanding of the distribution of observable hereditary characteristics in the decades before it was finally localized to the chromosome?

The demand that theoretical terms be linked to observations in ways that make a difference for predictions is far too strong a requirement; some theoretical terms, especially new ones, will not pass this test. It is also too weak a requirement, for it is easy to "cook up" a theory in which purely fictitious entities – vital forces, for example – play an indispensable role in the derivation of generalizations about what we can observe. If partial interpretation is too weak, we need to rethink the whole approach to what makes the unobservable terms of our theories meaningful and makes true or well-justified or even coherent the claims of science that the unobservable things these terms name actually exist.

But it may strike you that there is something about the way that logical empiricists treated this whole problem of the meaning of theoretical terms and the extent of our theoretical knowledge that gives it an artificial air. After all, though we may not be able to hear, taste, smell, touch or see electrons, genes, quasars and neutron stars, or their properties, we have every reason to think that they exist. For our scientific theories tell us that they do, and these theories have great predictive and explanatory power. If the most well-confirmed theory of the nature of matter includes the laws about molecules, atoms, leptons, bosons and quarks, then surely such things exist. If our most well-confirmed theories attribute charge, angular momentum, spin or van der Waals forces to these things, then surely such properties exist. On this view theories must be interpreted literally, not as making claims whose meaning is connected to observations, but as telling us about things and their properties, where the meaning of the names for these things and their properties is no more or less problematical than the meaning of terms that name observable things and their properties. And if this conclusion is incompatible with the theory of language enunciated above, which makes observational terms the basement level of language and requires all other terms to be built out of them, then so much the worse for that theory of language. And so much the worse for the empiricist epistemology that goes along with it.

This approach to the problem of theoretical terms is widely known as "**scientific realism**", since it takes the theoretical commitments of science to be real, and not just (disguised) abbreviations for observational claims, or useful fictions we create to organize these observations. Whereas the logical empiricist's starting point is a philosophical theory – empiricist epistemology, the scientific realist, or "realist" for short, starts with what realism takes to be a manifestly obvious fact about science: its great and ever-increasing predictive power. Over time our theories have improved both in the range and the precision of their predictions. Not only can we predict the occurrence of more and more different kinds of phenomena, but over time we have been able to increase the precision of our predictions – the number of decimal places or significant digits to which our scientifically derived expectations match up with our actual meter readings. These long-term improvements translate themselves into technological applications on which we increasingly rely, indeed on which we literally stake our lives every day. This so-called "instrumental success" of science cries out for explanation. Or at least the realist insists that it does. How can it be explained? What is the best explanation for the fact that science "works"? The answer seems evident to the realist: science works so well because it is (approximately) true. It would a miracle of cosmic proportions if science's predictive success and its technological applications were just lucky guesses, if science worked, as it were, by accident.

The structure of the scientific realist's argument is usually of the form:

1 **P**
2 The best explanation of the fact that **P**, is that **Q** is true.

Therefore,
3 **Q** is true.

Realists variously substitute for **P** the statement that science is predictively successful, or increasingly so, or that its technological applications are more and more powerful and reliable. For **Q** they substitute the statement that the unobservable things scientific theories postulate exist and have the properties science attributes to them; or else the realist makes a somewhat weaker claim like "Something like the unobservable entities that science postulates exist and have something like the properties that science attributes to them, and science is ever-increasing in its degree of approximation to the truth about these things and their properties." The structure of the argument from the truth of **P** to the truth of **Q** is that of an "**inference to the best explanation**".

This argument may strike the reader as uncontroversially convincing. It certainly appeals to many scientists. For they will themselves recognize that the inference-to-the-best-explanation form of reasoning the scientific realist philosopher uses is one they employ in science. For example, how do we know there are electrons and they have negative charges? Because postulating them explains the results of the Millikan Oil Drop Experiment and the tracks in a Wilson Cloud Chamber.

But the fact that the argument-form is used by science as well as used to justify science is its Achilles' heel. Suppose one challenges the argument to realism by demanding a justification for the inference-form given in 1–3 above. The realist's argument aims to establish scientific theorizing as literally true or increasingly approximate to the truth. If the realist argues that the inference form is reliable because it has been used with success in science, the realist's argument is potentially question-begging. In effect the realist argues that an inference to the best explanation's conclusion that scientific theorizing produces truths is warranted because science produces truths by using the inference-form in question. To use an analogy from the problem of induction in Chapter 3, this is rather like backing up a promise to return a loan by promising to keep the promise to repay.

What is more, the history of science teaches us that many successful scientific theories have completely failed to substantiate the scientific realist's picture of why theories succeed. Well before Kepler, and certainly since his time, scientific theories have not only been false and improvable, but if current science is any guide, they have sometimes been radically false in their claims about what exists and what the properties of things are, even as their predictive power has been persistently improved. One classical example is eighteenth-century phlogiston theory, which embodied significant predictive improvements over prior theories of combustion, but whose central explanatory entity, phlogiston, is nowadays cited with ridicule. Still another example is Fresnel's theory of light as a wave-phenomenon. This theory managed substantially to increase our predictive (and our explanatory) grasp on light and its properties. Yet the theory claims that light moves through a medium of propagation, an ether. The postulation of this ether is something one would expect in view of the difficulties traced above for the concept of gravity. Gravity is a mysterious force just because it doesn't seem to require any material through which to be transmitted. Without a medium of propagation, light would turn out to be as suspicious a phenomenon as gravity to the mechanistic materialism of nineteenth-century physics. Subsequent physics revealed that despite its great predictive improvements, the central theoretical postulate of Fresnel's theory, the ether, does not exist. It is not required by more adequate accounts of the behavior of light. Postulating the ether contributed to the "unrealism" of Fresnel's theory. This at least must be the judgment of contemporary scientific theory. But by a "pessimistic induction" from the falsity – sometimes radical falsity – of predictively successful theories in the past, it would be unsafe to assume that our current "best-estimate" theories are immune to a similar fate. Since science is fallible, one might expect that such stories can be multiplied to show that over the long term, as science progresses in predictive power and technological application, the posits of its theories vary so greatly in their reality as to undermine any straightforward inference to scientific realism's interpretation of its claims.

What is more, scientific realism is silent on how to reconcile the knowledge it claims we have about the (approximate) truth of our theories about unobservable entities with the empiricist epistemology that makes observation indispensable for knowledge. In a sense, scientific realism is part of the problem of how scientific knowledge is possible, not part of the solution.

One alternative to scientific realism much more sympathetic to empiricism has long attracted some philosophers and scientists. It bears the title "**instrumentalism**". This label names the view that scientific theories are useful instruments, heuristic devices, tools we employ for organizing our experience, but not literal claims about it that are either true or false. This philosophy of science goes back at least to the eighteenth-century British empiricist philosopher Berkeley, and is also attributed to leading figures of the Inquisition who sought to reconcile Galileo's heretical claims about the motion of the earth round the sun with holy writ and papal pronouncements. According to some versions of the history these learned churchmen recognized that the heliocentric hypothesis was at least as powerful in prediction as Ptolemaic theories, according to which the sun and the planets moved around the earth; they accepted that it might be simpler to use in calculations of the apparent positions of the planets in the night sky. But the alleged motion of the earth was observationally undetectable – it does not feel to us that the earth is moving. Galileo's theory required that we disregard the evidence of observation, or heavily reinterpret it. Therefore, these officers of the Inquisition urged Galileo to advocate his improved theory not as literally true, but as more useful, convenient and effective an instrument for astronomical expectations than the traditional theory. Were he so to treat his theory, and remain silent on whether he believed it was true, Galileo was promised that he would escape the wrath of the papal Inquisition. Although at first he recanted, Galileo eventually declined to adopt an instrumentalist view of the heliocentric hypothesis and spent the rest of his life under house arrest. Subsequent instrumentalist philosophers and historians of science have suggested that the Church's view was more reasonable than

Galileo's. And although Berkeley did not take sides in this matter, his arguments from the nature of language (sketched above) to the unintelligibility of realism (and of realistic interpretations of parts of Newton's theories), made instrumentalism more attractive. Berkeley went on to insist that the function of scientific theorizing was not to explain but simply to organize our experiences in convenient packages. On this view, theoretical terms are not abbreviations for observational ones, they are more like mnemonic devices, acronyms, uninterpreted symbols without empirical or literal meaning. And the aim of science is constantly to improve the reliability of its instruments, without worrying about whether reality corresponds to these instruments when interpreted literally.

It is worth noting that the history of the physical sciences from Newton onward shows a cyclical pattern of succession between realism and instrumentalism among scientists themselves. The realism of the seventeenth century, the period in which mechanism, corpuscularism and atomism held sway, was succeeded in the eighteenth century by the ascendancy of instrumentalist approaches to science, motivated in part by the convenient way with which instrumentalism dealt with Newton's mysterious force of gravity. By treating his theory of gravity as merely a useful instrument for calculating the motion of bodies, it could ignore the question of what gravity really is. By the nineteenth century, with advances in atomic chemistry, electricity and magnetism, the postulation of unobservable entities returned to favor among scientists. But then it again became unfashionable in the early twentieth century as problems for the realist's interpretation of quantum mechanics as a literally true description of the world began to mount. On the standard understanding of quantum mechanics, electrons and photons seem to have incompatible properties – being both wave-like and particle-like at the same time – and neither seem to have physical location until observed by us. These are two reasons why it is more than tempting to treat quantum mechanics as a useful instrument for organizing our experience in the atomic physics lab, and not as a set of claims true about the world independent of our observation of the world.

How does instrumentalism respond to the realists' claim that only realism can explain the instrumental success of science? The instrumentalist replies quite consistently with the following argument: that any explanation of the success of science that appeals to the truth of its theoretical claims either advances our predictive powers with respect to experience or it does not. If it does not, then we may neglect it and the question it purports to answer as without scientific, i.e. empirical, significance. If on the other hand, such an explanation would enhance the usefulness of our scientific instruments in systematizing and predicting experience, then instrumentalism can accept the explanation as confirming its treatment of theories as useful instruments instead of descriptions of nature.

There is a sort of halfway house between instrumentalism and realism worth briefly exploring. It is an attempt to have one's cake and eat it too: we agree with the scientist that scientific theories do purport to make claims about the world and especially about the unobservable underlying mechanisms which explain observations, and we can agree with the instrumentalist that knowledge of such claims is impossible. But we may argue that the objective of science should be or in fact is nothing more or less than systematizing experience. Therefore we can be agnostic about whether scientific theories are true, approximately true, false, convenient fictions or whatever. Just so long as they enable us to control and predict phenomena, we can and should accept them, without of course believing them (that would be to take a position on their truth). Science should be content with simply predicting with increasing precision and ever-wider range our experiences. In short, scientists should aim at what the instrumentalist recommends without embracing the instrumentalist's reason for doing so. It's not that science is an instrument. It's just that we cannot tell whether it is more than an instrument. And for all purposes it is enough that scientific theory be "empirically adequate". Recalling the words of the seventeenth-century natural philosophers, on this view, all we should demand of science is that it should "save the phenomena".

This combination of a realist interpretation for the claims of theoretical science with an instrumentalist epistemology has been called "**constructive empiricism**" by its developer, Bas van Fraassen. Few philosophers and fewer scientists will consider constructive empiricism to be an enduring stable equilibrium in the philosophy of science. After all, if science is either (increasingly approximately) true or (persistently) false in its representation of the world, although we can never tell which, then the treatment of science as a description of reality just drops out of intellectual matters. If we cannot tell which of these exhaustive and exclusive alternatives applies, then whichever does is probably irrelevant. And on the other hand, if we must forever withhold our judgment about the truth of the most predictively powerful and technologically successful body of hypotheses we can formulate, then the epistemological question of whether we can have scientific knowledge becomes as irrelevant to science as the skeptic's question of whether I am now dreaming or not.

Both realism and instrumentalism approach the problem of theoretical entities and the terms that name them with the same two assumptions in common. They are predicated on the assumption that we can distinguish the terms in which scientific laws and theories are expressed into observational ones and non-observational or theoretical ones; both agree that it is our knowledge of the behavior of observable things and their properties which tests, confirms and disconfirms our theories. For both, the court of last epistemological resort is observation. And yet, as we shall see below, how observation tests any part of science, theoretical or not, is no easy thing to understand.

3 Theories and models

Axiomatization is plainly not the way in which scientists actually present their theories. It does not pretend to be, seeking rather a rational reconstruction of the ideal or essential nature of a scientific theory which explains how it fulfils its function. But there are two immediate and related problems the axiomatic model faces. The first is that nowhere in the axiomatic account does the concept of model figure. And yet nothing is more characteristic of theoretical science than its reliance on the role of models. Consider the planetary model of the atom, the billiard-ball model of a gas, Mendelian models of genetic inheritance, the Keynesian macroeconomic model. Indeed, the very term "**model**" has supplanted the word "theory" in many contexts of scientific inquiry. It is pretty clear that often the use of this term suggests the sort of tentativeness that the expression "just a theory" conveys in non-scientific contexts. But in some domains of science there seem to be nothing but models, and either the models constitute the theory or there is no separate thing at all that is properly called a theory. This is a feature of science that the axiomatic approach must explain or explain away.

The second of our two problems for the axiomatic approach is the very idea that a theory is an axiomatized set of sentences in a formalized mathematical language. The claim that a theory is an axiomatic system is in immediate trouble, in part because, as we noted above, there are many different ways to axiomatize the same set of statements. But more than that, an axiomatization is essentially a linguistic thing: it is stated in a particular language, with a particular vocabulary of defined and undefined terms, and a particular syntax or grammar. Now ask yourself, is Euclidean geometry correctly axiomatized in Greek, with its alphabet, or German with its gothic letters, its verbs at the end of sentences and its nouns inflected, or in English? The answer is that Euclidean geometry is indifferently axiomatized in any language in part because it is not a set of sentences in a language but a set of propositions which can be expressed in an indefinite number of different axiomatizations in an equally large number of different languages. To confuse a theory with its axiomatization in a language is like confusing the number 2 – an abstract object – with the concrete inscriptions, like "dos", "II", "zwei", "$10_{(base\ 2)}$" we employ to name it. Confusing a theory with its axiomatization is like mistaking a proposition (again, an abstract object) for the particular sentence (a concrete object) in a language used to express it. "Es regnet" is no more the proposition that it is raining than "Il pleut", nor is "It's raining" the correct way to express the proposition. All three of these inscriptions express the same proposition about the weather, and the proposition itself is not in any language. Similarly, we may not want to identify a theory with its axiomatization in any particular language, not even in some perfect, mathematically powerful, logically clear language. And if we don't want to do this, the axiomatic account is in some difficulty, to say the least.

What is the alternative? Let's start with models for phenomena that scientists actually develop, for example, the Mendelian model of the gene. A Mendelian gene is any gene which assorts independently and segregates from its allele in meiosis. Notice that this statement is true by definition. It is what we mean by "Mendelian gene". Similarly, we may express the model for a Newtonian system: A Newtonian system is any set of bodies that behave in accordance with the following two formulae, $F = Gm_1m_2/d^2$ – the inverse square law of gravitation attraction – and $F = ma$ – the law of free-falling bodies – and with the laws of rectilinear motion and the conservation of momentum. Again, these four features define a Newtonian system. Now, let's consider what arrangement of things in the world satisfies these definitions? Well, by assuming that the planets and the sun are a Newtonian system, we can calculate the positions of all the planets with great accuracy as far into the future and as far into the past as we like. So, the solar system satisfies the definition of a Newtonian system. Similarly, we can calculate eclipses – solar and lunar – by making the same assumption for the sun, the earth and the moon. And of course we can do this for many more sets of things – cannonballs and the earth, inclined planes and balls, pendula. In fact, if we assume that gas molecules satisfy our definition of a Newtonian system, then we can predict their properties too.

Notice that the definition given above for a Newtonian system is not the only definition we could give. Following Richard Feynman, we may, for example, substitute for the inverse square formula one which relates the gravitational potential on an object at a point to the average gravitational potential surrounding that point: $\Phi = $ average $\Phi - Gm/2a$, where Φ is the gravitational potential at any given point, a is the radius of the surrounding sphere on the surface of which the average potential, average Φ, is calculated, G is the same constant as figures in the formula above and m is the mass of the objects at the point on which gravity is exerted. Feynman argued that one may prefer this formula to the usual one because $F = Gm_1m_2/d^2$ suggests that gravitational force operates over large distances instantaneously, whereas the less familiar equation gives the values of gravitational force at a point in terms of values at other points which can be as close as one arbitrarily chooses. But either definition will work to characterize a Newtonian gravitational system.

Now the reason we call these definitions models is that they "fit" some natural processes more accurately than others; that they are often deliberate simplifications which neglect causal variables we know exist but are small compared to the ones the model mentions; and that even when we know that things in the world don't really fit them at all, they may still be useful calculating devices, or pedagogically useful ways of introducing a subject. Thus, a Newtonian model of the solar system is a deliberate simplification which ignores friction, small bodies like comets, moons and asteroids, and electric fields, among other things. Indeed, we know that the model's exact applicability is disconfirmed by astronomical data on, for example, Mercury's orbit. And we know that the model's causal variable does not really exist (there is no such thing as Newtonian gravity which acts at a distance; rather space is curved). Nevertheless, it is still a good model for introducing mechanics to the student of physics and for sending satellites to the nearest planets. Moreover, the advance of mechanics from Galileo and Kepler to Newton and Einstein is a matter of the succession of models, each of which is applicable to a wider range of phenomena and/or more accurate in its predictions of the behavior of the phenomena.

A model is true by definition. An ideal gas is by definition just what behaves in accordance with the ideal-gas law. The empirical or factual question about a model is whether it "applies" to anything closely enough to be scientifically useful – to explain and predict its behavior. Thus, it will be a hypothesis that the Newtonian model applies well enough to, or is sufficiently well satisfied by, the solar system. Once we specify "well-enough" or "sufficiently well satisfied", this is a hypothesis that usually turns out to be true. The unqualified claim that the solar system is a Newtonian system is, we know, strictly speaking false. But it is much closer to the truth than any other hypothesis about the solar system except the hypothesis that the solar system satisfies the model propounded by Einstein in the general theory of relativity. And a theory? A theory is a set of hypotheses claiming that particular sets of things in the world are satisfied to varying degrees by a set of models which reflect some similarity or unity. This will usually be a set of successively more complex models. For example, the kinetic theory of gases is a set of models that begins with ideal-gas law we have seen before, $PV = rT$. This model treats molecules as billiard balls without intermolecular forces and assumes they are mathematical points. The theory includes a subsequent improvement due to van der Waals, $(P + a/V^2)(V - b) = rT$, in which a

represents the intermolecular forces and **b** reflects the volume molecules take up, both neglected by the ideal-gas law. And there are other models as well, Clausius's model, and ones that also introduce quantum considerations.

Exponents of this approach to theories, according to which they are sets of models, that is of formal definitions, along with claims about what things in the world salsify these definitions, call their analysis the "**semantic**" account of scientific theories and contrast it to the axiomatic account which they call the "syntactic" account for two related reasons: (a) it requires derivation of empirical generalizations from axioms in accordance with rules of logic, which are the syntax of the language in which the theory is stated; (b) the derivations which logical rules permit operate on the purely formal features – the syntax – of the axioms, and not the meaning of their terms. Notice that although models will be linguistic items on the semantic view – definitions – hypotheses and theories will not be linguistic items but (abstract) propositions expressible in any language, to the effect that the world or some part of it satisfies to some degree or other one or more models, expressed indifferently in any language convenient for doing so.

But surely this is not the chief advantage of the semantic view, by comparison to the syntactic view. For after all, the axiomatic account may well be best understood as the claim that a theory is a set of axiom systems in any language that expresses all the same propositions as axioms or theorems, or that it is the set of all such axiom systems that best balance simplicity and economy of expression with power in reporting these propositions. If the linguistic or non-linguistic character of theories is a problem, it is a rather technical one for philosophers, which should have little impact on our understanding of scientific theories. The advantage of semantical over syntactical approaches to theories must lie elsewhere.

One advantage the semantical approach has of course is that it focuses attention on the role and importance of models in science in a way that the axiomatic account does not. In particular, it is hard for the axiomatic analysis to accommodate the formulation of models known from the outset to be at most false but useful idealizations. It won't do to simply to interpret $PV = rT$ not as a definition of an ideal gas, but as an empirical generalization about real objects to be derived from axioms of the kinetic theory of gases, if we know that the statement is false and could not be true. We don't want to be able to derive such falsehoods directly from our axiomatic system. For such derivations imply that one or more of the axioms is false. What we may want is to find a place for models within an axiomatic approach.

A related advantage of the semantic approach is often claimed for it. In

some areas of science, it is sometimes claimed that there is no axiomatization available of the relevant laws, or that axiomatization would be
premature and freeze the development of ideas which are still being
formulated. To suggest that thinking in a discipline can or should be
rationally reconstructed as an axiomatization would therefore be disadvantageous. Sometimes it is claimed that evolutionary theory in biology
is like this, still too fluid a subject to be formalized into one canonical
expression of its contents. The many Mendelian and post-Mendelian
models in population biology (almost all of them definitions in the sense
of the semantic approach) are packaged together as evolutionary theory.
But, according to many philosophers of biology, there is not yet anything
more general in the theory than these models. When we try to frame the
theory of natural selection into an axiomatic system, the result is often
rejected by evolutionary biologists as failing to adequately reflect the full
richness of Darwin's theory and its latter-day extensions. In particular, the
theory's assertion that the fittest among competing organisms survive
and reproduce is one easy to deprive of its explanatory force if we define
"the fittest" as those which survive and reproduce. But it is hard to see
how to define fitness in a way that does not trivialize any axiomatization
of the theory. Moreover, there seems to be no agreed-upon evolutionary
theoretical structure in which to embed the genetic models which evolutionary biologists formulate and apply. All these are reasons that
philosophers of biology have accepted the semantic view as capturing
more adequately the character of theory in biology.

But can particular sciences or subdisciplines like evolutionary biology,
or even compartments of disciplines that go by the name of "theory",
such as evolutionary theory, game theory, economic theory, cognitive
theory, really remain agnostic about the existence of fundamental underlying theories towards which models in their disciplines are moving?
They must do so, if there simply is no set of higher-level general laws in
the discipline that explains lower-level regularities, and their exceptions.

Recall one of the metaphysical attractions of the axiomatic approach:
its commitment to axiomatization as an account of how a theory explains
by uncovering underlying mechanisms. Consider the metaphysical thesis
that at bottom the universe is simple in composition and operation, and
all the diversity of more complicated and more composite things is the
result of the simplicity at the bottom of things. This thesis suggests that
there is a true theory about layers of causal laws, each of which rests on a
more fundamental layer of smaller numbers of laws about a smaller range
of simpler objects that imply the less fundamental laws. It is a short step
to the conclusion that there should be a uniquely correct axiomatization
of this theory that reflects the structure of reality. The logical empiricists
who first advanced the axiomatic account would not have expressed such

a view because of their desire to avoid controversial metaphysical debate. Philosophers less averse to metaphysics will certainly find the view a motivation for adopting a syntactic model of theories. By contrast, philosophers who reject this metaphysical picture have a concomitant reason to adopt the semantic approach to theories. For this approach makes no commitments to any underlying simplicity or to the reducibility of less fundamental theories (i.e. sets of models) to more fundamental theories (i.e. sets of more fundamental models). If nature is just not simple, the structure of science will reflect this fact in a plethora of sets of models, and a dearth of axiomatic systems. And it will encourage instrumentalism about the character of theories and their claims about reality.

Notice that the instrumentalist can refuse even to be a party to this debate about whether theories describe reality. For the instrumentalist must be indifferent to the question of whether there is some set of laws which explain why the models work. Indeed, so far as instrumentalism is concerned, models might just as well supplant theory altogether in the advancement of science. Who needs theory if it cannot provide greater empirical adequacy than the models whose success it explains. It is for this reason that it is sometimes supposed that the semantic view of theories is more amenable to an instrumentalist philosophy of science than the syntactic or axiomatic approach.

By contrast for the realist both the success and especially the increasing accuracy of the succession of models in these subdisciplines demands explanation. Of course, some may argue that it is possible for a set of models in, say, evolutionary biology, to provide considerable predictive power and indeed increasing precision, even though the only general theory in biology is to be found at the level of molecular biology. For example, it might turn out that the biological models we formulate work for creatures with our peculiar cognitive and computational limitations and our practical interests, but that the models don't really reflect the operation of real laws at the level of organization of organisms and populations of them. This would be a realist's explanation for the absence of laws at some levels of organization where there are effective models. But the realist cannot adopt such a stratagem to explain away the absence of laws that might explain the success of models in physics or chemistry.

Moreover, the realist will argue, the semantic approach shares with the axiomatic account a commitment to the existence of theories distinct from and different from the models on which it focuses. For the semantic approach tells us that a theory is the substantive claim that a set of models which share some features in common are satisfied by things in the world. A theory is the set of definitions that constitute the models *plus* the claim that there are things that realize, satisfy, instantiate, exemplify these definitions sufficiently well to enable us to predict their behavior (observable or unobservable) to some degree of accuracy. Applying a model to real processes is an *ipso facto* commitment to the truth of this substantive claim. But such a claim is more than a mere instrument or useful tool that enables us to organize our experiences. Accordingly, like the axiomatic account, the semantic approach is committed to the truth of general claims in science. And the semantic view of theories has all the same intellectual obligations as the axiomatic account does, to explain why theories are true or approximately true or at least moving successively closer to the truth.

Moreover, the semantic view of theories faces the same problems as those with which we left the axiomatic account at the end of the last section. Since many of the models in science are definitions of unobserved, theoretical systems, such as the Bohr model of the atom to take a century-old example, the semantic view of theories faces the same problem of reconciling empiricism with the indispensability of theoretical terms, or equivalently the commitment to theoretical objects as the axiomatic account does. Applying a model to the world requires that we connect it to what can be observed or experienced, even if what is observed is a photograph that we interpret as representing a subatomic collision, or a binary star or the semi-conservative replication of a DNA molecule. Whether the theory (or a model) explains data as the realist holds, or only organizes it as the instrumentalist holds, the theory can do neither without recourse to claims about this realm of unobservable things, events, processes, properties that an empiricist epistemology makes problematic. But the final epistemic arbiter for science is observation. And yet, as we shall see in Chapter 5, how observation tests any part of science, theoretical or not, is no easy thing to understand.

Summary

The axiomatic account of scientific theories explains how the theoretical laws of a theory work together to provide an explanation of a large number of empirical or observable regularities by treating theories as deductively organized systems, in which the assumptions are hypotheses confirmed by the observations that confirm the generalization derived from them. This conception of laws as hypotheses tested by the consequences deduced from them is known as "hypothetico-deductivism", a well-established account of how theories and experience are brought together.

Theories often explain by identifying the underlying unobserved processes or mechanisms that bring about the observable phenomena which test the theories. Reductionism labels a long-standing view about the relationship of scientific theories to one another. According to reductionism, as a science deepens its understanding of the world, narrower, less accurate and more special theories are revealed to be special cases of or explainable by derivation from broader, more complete, more accurate and more general theories. Derivation requires the logical deduction of the axioms of the narrower theory from the broader theory, and often the correction of the narrower theory before the deduction is effected. Reductionists seek to explain the progress of science over the period since the Newtonian revolution by appeal to these intertheoretical relations. The reduction of scientific theories over centuries, which seems to preserve their successes while explaining their failures (through correction), is easy to understand from the axiomatic perspective on the structure of scientific theories.

However, the hypothetico-deductivism of the axiomatic account of theories, and indeed the general epistemological perspective of science as based on observation and experiment, faces grave difficulty when it attempts to explain the indispensability of terms in theories that identify theoretical, unobservable entities, like cellular nuclei, genes, molecules, atoms and quarks. For on the one hand, there is no direct evidence for the existence of the theoretical entities these terms name, and on the other hand, theory cannot discharge its explanatory function without them. Some theoretical entities, such as gravitation, are truly troublesome, and at the same time, we need to exclude from science mysterious and occult forces and things for which no empirical evidence can be provided. The notion that meaningful words must eventually have their meanings given by experience is an attractive one. Yet finding a way for theoretical language to pass this test, while excluding the terms of uncontrolled speculation as meaningless, is a challenge that an account of scientific theories must face.

The puzzle that hypothesizing theoretical entities is indispensable to explanation and unregulated by experience is sometimes solved by denying that scientific theories seek to describe the underlying realities that systematize and explain observational generalizations. This view, known as instrumentalism, or antirealism, treats theory as a heuristic device, a calculating instrument for predictions alone. By contrast, realism, the view that we should treat scientific theory as a set of literally true or false descriptions of unobservable phenomena, insists that only the conclusion that theory is approximately true can explain its long-term predictive success. Instrumentalists controvert this explanation.

The axiomatic approach to theories has difficultly accommodating the role of models in science. Instrumentalism does not, and as models become more central to the character of scientific theorizing, problems for the axiomatic approach and for realism mount. The issue here ultimately turns on whether science shows a pattern of explanatory and predictive successes which can only be explained by realism and the existence of theories that organize and explain the success of the models scientists develop.

Questions

1 Deductive or axiomatic systems do not seem to provide an illuminating account of how the components of a theory "work together". After all, any two laws can figure as the axioms of some theory or other made up on the spur of the moment. Can you offer a more precise notion of how the laws in a theory "work together"?
2 Is "constructive empiricism" really a viable middle course between instrumentalism and realism?
3 Evaluate the following argument for realism: "As technology progresses, yesterday's theoretical entities become today's observable ones. Nowadays we can detect cells, genes and molecules. In the future we will be able to observe photons, quarks, etc. This will vindicate realism."
4 What makes the semantic approach, with its emphasis on models, more amenable to instrumentalism than to realism?
5 Does instrumentalism owe us an explanation of the success of science? If so, what is it? If not, why not?

Further reading

The history of philosophical analysis of scientific theorizing is reported in F. Suppe, *The Structure of Scientific Theories*. The axiomatic approach was perhaps first fully articulated in R. Braithwaite, *Scientific Explanation*. Perhaps the most influential and extensive account of theories, and of science in general, to emerge from the period of logical empiricism is E. Nagel, *The Structure of Science*, first published in 1961. This magisterial work is worthy of careful study on all topics in the philosophy of science. Its account of the nature of theories, its development of examples and its identification of philosophical issues remains unrivaled. Nagel's discussion of the structure of theories, of reductionism and of the realism/antirealism issue set the agenda for the next several decades. The view of scientific progress reflected in Nagel's notion of reduction is examined in W. Newton-Smith, *The Rationality of Science*. M. Spector, *Concepts of Reduction in Physical Science*, and A. Rosenberg, *The Structure of Biological Science*, expound and examine the relations among theories in these two compartments of natural science. But many papers have been written and continue to appear on this issue, especially in the journals *Philosophy of Science* and *The British Journal for Philosophy of Science*.

Hempel's paper, "The Theoreticians Dilemma", in *Aspects of Scientific Explanation*, expresses the problem of reconciling the indispensability of theoretical entities for explanation with the empiricist demand that the terms naming those entities be observationally meaningful. Other papers in *Aspects*, including "Empiricist Criteria of Significance: Problems and Changes", reflect these problems. Among the earliest and most vigorous post-positivist arguments for realism is J. J. C. Smart, *Between Science and Philosophy*. The debate between realists and antirealists or instrumentalists to which Hempel's problem gives rise is well treated in J. Leplin (ed.), *Scientific Realism*, which includes papers defending realism by R. Boyd, and E. McMullin; a development of the "pessimistic induction" from the history of science to the denial of realism by L. Laudan; a statement of van Fraassen's "constructive empiricism"; and a plague on both realism and antirealism pronounced by Arthur Fine, "The Natural Ontological Attitude". Van Fraassen's views are more fully worked out in *The Scientific Image*. J. Leplin, *A Novel Argument for Scientific Realism*, is a more recent defense of realism against van Fraaseen and others. P. Churchland and C. A. Hooker (eds), *Images of Science: Essays on Realism and Empiricism*, is a collection of essays discussing "constructive empiricism".

The semantic view of theories is elaborated by F. Suppe in *The Structure of Scientific Theories* as well as by van Fraassen, *The Scientific Image*. Its application to biology is treated in P. Thompson, *The Structure of Biological Theories*, and E. Lloyd, *The Strucutre of Evolutionary Theory*.

CHAPTER 5
The epistemology of scientific theorizing

Overview

S uppose we settle the dispute between realism and instrumentalism. The problem still remains of how exactly observation and evidence, the collection of data, etc., actually enable us to choose among scientific theories. On the one hand, that they do so has been taken for granted across several centuries of science and its philosophy. On the other hand, no one has fully explained how they do so, and in the twentieth century the challenges facing the explanation of exactly how evidence controls theory have increased.

A brief review of the history of British empiricism sets the agenda for an account of how science produces knowledge justified by experience. Even if we can solve the problem of induction raised by Hume, or show that it is a pseudo-problem, we must face the question of what counts as evidence in favor of a hypothesis. The question seems easy, but it turns out to be a very complex one on which the philosophy of science has shed much light without answering to everyone's satisfaction.

Modern science makes great use of statistical methods in the testing of hypotheses. We explore the degree to which a similar appeal to probability theory on behalf of philosophy can be used adequately to express the way data support theory. Just as the invocation of probability in Chapter 2 leads to questions of how we are to understand this notion, invoking it to explain confirmation of hypotheses forces us to choose among alternative interpretations of probability.

Even if we adopt the most widely accepted account of theory confirmation, we face a further challenge: the thesis of underdetermination, according to which even when all the data are in, the data will not by themselves choose among competing scientific theories. Which theory, if any, is the true theory may be underdetermined by the evidence even when all the evidence is in. This conclusion, to the extent it is adopted, not only threatens the empiricist's picture of how knowledge is certified in science but threatens the whole edifice of scientific objectivity altogether, as Chapter 6 describes.

1 A brief history of empiricism as science's epistemology

The scientific revolution began in central Europe with Copernicus, Brahe and Kepler, shifted to Galileo's Italy, moved to Descartes' France and ended with Newton in Cambridge, England. The scientific revolution was also a philosophical revolution, and for reasons we have already noted. In the seventeenth century science was "natural philosophy", and figures that history would consign exclusively to one or the other of these fields contributed to both. Thus Newton wrote a good deal of philosophy of science, and Descartes made contributions to physics. But it was the British empiricists who made a self-conscious attempt to examine whether the theory of knowledge espoused by these scientists would vindicate the methods which Newton, Boyle, Harvey and other experimental scientists employed to expand the frontiers of human knowledge so vastly in their time.

Over a period from the late seventeenth century to the late eighteenth century, John Locke, George Berkeley and David Hume sought to specify the nature, extent and justification of knowledge as founded on sensory experience and to consider whether it would certify the scientific discoveries of their time as knowledge and insulate them against skepticism. Their results were mixed, but nothing would shake their confidence or that of most scientists, in empiricism as the right epistemology.

Locke sought to develop empiricism about knowledge, famously holding against rationalists like Descartes, that there are no innate ideas. "Nothing is in the mind that was not first in the senses." But Locke was resolutely a realist about the theoretical entities which seventeenth-century science was uncovering. He embraced the view that matter was composed of indiscernible atoms, "corpuscles" in the argot of the time, and distinguished between material substance and its properties on the one hand, and the sensory qualities of color, texture, smell or taste, which it causes in us. The real properties of matter, according to Locke, are just the ones that Newtonian mechanics tells us it has – mass, extension in space, velocity, etc. The sensory qualities of things are ideas in our heads which the things cause. It is by reasoning back from sensory effects to physical causes that we acquire knowledge of the world, which gets systematized by science.

That Locke's realism and his empiricism inevitably gives rise to skepticism, is not something Locke recognized. It was a philosopher of the next generation, George Berkeley, who appreciated that empiricism makes doubtful our beliefs about things we do not directly observe. How could Locke lay claim to the certain knowledge of the existence of matter or its features, if he could only be aware of sensory qualities, which by their very nature exist only in the mind. We cannot compare sensory features like color or texture to their causes to see whether these causes are color-less or not, for we have no access to these things, so we cannot compare them. And to the argument that we can imagine something to be color-less, but we cannot imagine a material object to lack extension or mass, Berkeley retorted that sensory properties and non-sensory ones are on a par in this respect: try to imagine something without color. If you think of it as transparent, then you are adding in the background color and that's cheating. Similarly for the other allegedly subjective qualities that things cause us to experience.

In Berkeley's view, without empiricism we cannot make sense of the meaningfulness of language. Berkeley pretty much adopted the theory of language as naming sensory qualities that was sketched in the last chapter. Given the thesis that words name sensory ideas, realism – the thesis that science discovers truths about things we cannot have sensory experience of – becomes false, for the words that name these things must be meaningless. In place of realism Berkeley advocated a strong form of instrumentalism and took great pains to construct an interpretation of seventeenth- and eighteenth-century science, including Newtonian mechanics, as a body of heuristic devices, calculating rules and convenient fictions we employ to organize our experiences. Doing this, Berkeley thought, saves science from skepticism. It did not occur to Berkeley that another alternative to the combination of empiricism and instrumen-talism is rationalism and realism. And the reason is that, by the eighteenth century, the role of experiment in science had been so securely established that no alternative to empiricism seemed remotely plausible as an epistemology for science.

Indeed, it was David Hume's intention to apply what he took to be the empirical methods of scientific inquiry to philosophy. Like Locke and Berkeley, he sought to show how knowledge, and especially scientific knowledge, honors the strictures of empiricism. Unable to adopt Berkeley's radical instrumentalism, Hume sought to explain why we adopt a realistic interpretation of science and ordinary beliefs, without taking sides between realism and instrumentalism. But, as we saw in Chapter 2, Hume's pursuit of the program of empiricism led him to face a problem different from that raised by the conflict of realism and empiri-cism. This is the problem of induction: given our current sensory experience, how can we justify inferences from them and from our records of the past, to the future and to the sorts of scientific laws and theories we seek.

Hume's argument is often reconstructed as follows: there are two and only two ways to justify a conclusion: deductive argument, in which the conclusion follows logically from the premises, and inductive argument, in which the premises support the conclusion but do not guarantee it. A deductive argument is colloquially described as one in which the premises "contain" the conclusion, whereas an inductive argument is often described as one that moves from the particular to the general, as when we infer from observation of 100 white swans to the conclusion that all swans are white. Now, if we are challenged to justify the claim that inductive arguments – arguments from the particular to the general, or from the past to the future – will be reliable in the future, we can do so only by employing a deductive argument or an inductive argument. The trouble with any deductive argument to this conclusion is that at least one of the premises will itself require the reliability of induction. For example, consider the deductive argument below:

1 If a practice has been reliable in the past, it will be reliable in the future.
2 In the past inductive arguments have been reliable.

Therefore
3 Inductive arguments will be reliable in the future.

This argument is deductively valid, but its first premise requires justification, and the only satisfactory justification for the premise would be the reliability of induction, which is what the argument is supposed to establish. Any deductive argument for the reliability of induction will include at least one question-begging premise. This leaves only inductive arguments to justify induction. But clearly, no inductive argument for induction will support its reliability, for such arguments too are question-begging. As we have had occasion to note before, like all such question-begging arguments, an inductive argument for the reliability of induction is like underwriting your promise to pay back a loan by promising that you keep your promises. If your reliability as a promise-keeper is what is in question, offering a second promise to assure the first one is pointless. Hume's argument has for 250 years been treated as an argument for skepticism about empirical science, for it suggests that all conclusions about scientific laws, and all prediction science makes about future events, are at bottom unwarranted, owing to their reliance on induction. Hume's own conclusion was quite different. He noted that as a person who acts in the world, he was satisfied that inductive arguments were reasonable; what he thought the argument shows is that we have not yet found the right justification for induction, not that there is no justification for it.

The subsequent history of empiricism shares Hume's belief that there is a justification for induction, for empiricism seeks to vindicate empirical science as knowledge. Throughout the nineteenth century philosophers like John Stuart Mill sought solutions to Hume's problem. In the twentieth century many logical positivists too believed that a solution could be found for the problem of induction. One such positivist argument (due to Hans Reichenbach) seeks to show that if any method of predicting the future works, then induction must work. Suppose we wish to establish whether the oracle at Delphi is an accurate predictive device. The only way to do so is to subject the oracle to a set of tests: ask for a series of predictions and determine whether they are verified. If they are, the oracle can be accepted as an accurate predictor. If not, then the future accuracy of the oracle is not to be relied upon. But notice that the form of this argument is inductive. If any method works (in the past), only induction can tell us that it does (in the future). Whence we secure the justification of induction. This argument faces two difficulties. First, at most it proves that if any method works, induction works. But this is a far cry from the conclusion we want: that any method does in fact work. Second, the argument will not sway the devotee of the oracle. Oracle-believers will have no reason to accept our argument. They will ask the oracle whether induction works, and will accept its pronouncement. No attempt to convince oracle-believers that induction supports either their method of telling the future or any other can carry any weight with them. The argument that if any method works, induction works, is question-begging too.

Other positivists believed that the solution to Hume's problem lay in disambiguating various notions of probability, and applying the results of a century's advance in mathematical logic to Hume's empiricism. Once the various senses of probability employed in science were teased apart, they hoped either to identify the notion that is employed in scientific reasoning from data to hypotheses, or to explicate that notion to provide a "rational reconstruction" of scientific inference that vindicates it. Recall the strategy of explicating scientific explanation as the D-N model. The positivists spent more time attempting to understand and explicate the logic of the experimental method – inferring from data to hypotheses – than on any other project in the philosophy of science. The reason is obvious. Nothing is more essential to science than learning from experience, that is what is meant by empiricism. And they believed this was the way to find a solution to Hume's problem.

Some of what Chapter 3 reports about interpretations of probability reflects the work of these philosophers. In this chapter we will encounter more of what they uncovered about probability. What these philosophers and their students discovered about the logical foundations of probability, and of the experimental method in general, turned out to raise new problems beyond those which Hume laid before his fellow empiricists.

2 The epistemology of scientific testing

There is a great deal of science to do long before science is forced to invoke unobservable things, forces, properties, functions, capacities and dispositions to explain the behavior of things observable in experience and the lab. Even before we infer the existence of theoretical entities and processes, we are theorizing. A scientific law, even one exclusively about what we can observe, goes beyond the data available, because it makes a claim which if true is true everywhere and always, not just in the experience of the scientist who formulates the scientific law. This of course makes science fallible: the scientific law, our current best-estimate hypothesis, may turn out to be, in fact usually does turn out to be, wrong. But it is by experiment that we discover this, and by experiment that we improve on it, presumably getting closer to the natural law we seek to discover.

It may seem a simple matter to state the logical relationship between the evidence that scientists amass and the hypotheses the evidence tests. But philosophers of science have discovered that testing hypotheses is by no means an easily understood matter. From the outset it was recognized that no general hypothesis of the form "All **A**s are **B**s" – for instance, "All samples of copper are electrical conductors" – could be conclusively confirmed, because the hypothesis will be about an indefinite number of **A**s and experience can provide evidence only about a finite number of them. By itself a finite number of observations, even a very large number, might be only an infinitesimally small amount of evidence for a hypothesis about a potentially infinite number of, say, samples of copper. At most, empirical evidence supports a hypothesis to some degree. But as we shall see, it may also support many other hypotheses to an equal degree.

On the other hand, it may seem that such hypotheses could at least be falsified. After all, to show that all **A**s are **B**s is false, one need only find an **A** which is not a **B**: after all, one black swan refutes the claim that all swans are white. And understanding the logic of **falsification** is particularly important because science is fallible. Science progresses by subjecting a hypothesis to increasingly stringent tests, until the hypothesis is falsified, so that it may be corrected, improved, or give way to a better hypothesis. Science's increasing approximation to the truth relies crucially on falsifying tests and scientists' responses to them. Indeed, some philosophers of science have argued that scientists never seek evidence to confirm hypotheses at all, but only to falsify them. Following Karl Popper, these philosophers argue that since no finite amount of evidence could ever distinguish between the vast number of possible hypotheses scientists can come up with, scientists can't be in the business of confirming hypotheses at all. But they don't need to either. Scientists seek falsification of their claims, not confirmation. For it is only by falsifying hypotheses that we can learn how to improve their predictive accuracy and explanatory power. For these philosophers, the empirical content of a hypothesis is a matter of whether there are possible observations that would show it to be false.

But the claim that general hypotheses are strictly falsifiable is not correct. For nothing follows from a general law alone. From "All swans are white" it does not follow that there are any swans, still less that there are white ones. To test this generalization we need to independently establish that there is at least one swan and then check its color. Testing even the simplest hypothesis requires "auxiliary assumptions" – further statements about the conditions under which the hypothesis is tested. For example to test "All swans are white" we need to establish that "this bird is a swan", and doing so requires we assume the truth of other generalizations about swans besides what their color is. What if the grey bird before us is a grey goose? No single falsifying test will tell us whether the fault lies with the hypothesis under test or with the auxiliary assumptions we need to uncover the falsifying evidence.

To see the problem more clearly consider a test of $PV = rT$. To subject the ideal-gas law to test we measure two of the three variables, say the volume of the gas container and temperature, use the law to calculate a predicted pressure, and then compare the predicted gas pressure to its actual value. If the predicted value is identical to the observed value, the evidence supports the hypothesis. If it does not, then presumably the hypothesis is falsified. But in this test of the ideal-gas law we needed to measure the volume of the gas and its temperature. Measuring its temperature requires a thermometer, and employing a thermometer requires us to accept one or more rather complex hypotheses about how thermometers measure heat, for example the scientific law that mercury in an enclosed glass tube expands as it is heated, and does so uniformly. But this is another general hypothesis – an auxiliary we need to invoke in order to put the ideal-gas law to the test. If the predicted value of the pressure of the gas diverges from the observed value, the problem may be that our thermometer was defective, or that our hypothesis about how expansion of mercury in an enclosed tube measures temperature change is false. But to show that a thermometer was defective, because say the glass tube was broken, presupposes another general hypothesis: thermometers with broken tubes do not measure temperature accurately. Now in many cases of testing of course the auxiliary hypotheses are among the most basic generalizations of a discipline, like acid turns blue litmus paper red, which no one would seriously challenge. But the logical possibility that they might be mistaken, a possibility that cannot be denied, means that any hypothesis which is tested under the assumption that the auxiliary assumptions are true, can be in principle preserved from falsification, by giving up the auxiliary assumptions and attributing the falsity to these auxiliary assumptions. And sometimes, hypotheses are in practice preserved from falsification. Here is a classic example in which the falsification of a test is rightly attributed to the falsity of auxiliary hypotheses and not the theory under test. In the nineteenth century predictions of the location in the night sky of Jupiter and Saturn derived from Newtonian mechanics were falsified as telescopic observation improved. But instead of blaming the falsification on Newton's laws of motion, astronomers challenged the auxiliary assumption that there were no other forces, beyond those due to the known planets, acting on Saturn and Neptune. By calculating how much additional gravitational force was necessary and from what direction, to render Newton's laws consistent with the data apparently falsifying them, astronomers were led to the discovery, successively, of Neptune and Uranus.

As a matter of logic, scientific law can neither be completely established by available evidence, nor conclusively falsified by a finite body of evidence. This does not mean that scientists are not justified on the occasions at which they surrender hypotheses because of countervailing evidence, or accept them because of the outcome of an experiment. What it means is that confirmation and disconfirmation are more complex matters than the mere derivation of positive or negative instances of a hypothesis being tested. Indeed, the very notion of a positive instance turns out to be a hard one to understand.

Consider the hypothesis that "all swans are white". Here is a white bird which is a swan, and a black boot. Which is a positive instance of our hypothesis? Well, we want to say that only the white bird is; the black boot has nothing to do with our hypothesis. But logically speaking, we have no right to draw this conclusion. For logic tells us that "all **As** are **Bs**" if and only if "all non-**Bs** are non-**As**". To see this, consider what would be an exception to "all **As** are **Bs**"? It would be an **A** that was not a **B**. But this would also be an exception to "All non-**Bs** are non-**As**". Accordingly, statements of these two forms are logically equivalent. In consequence, all swans are white if and only if all non-white things are non-swans. The two sentences are logically equivalent formulations of the same statement. Since the black boot is a non-white non-swan, it is a positive instance of the hypothesis that all non-white things are non-swans, aka all swans are white. The black boot is a positive instance of the hypothesis that all swans are white. Something has gone seriously wrong here! Surely the way to assess a hypothesis about swans is not to examine boots! At a minimum, this result shows that the apparently simple notion of a "positive instance" of a hypothesis is not so simple, and one we do not yet fully understand.

One conclusion drawn from the difficulty of this problem supports Popper's notion that scientists don't or at least shouldn't try to confirm hypotheses by piling up positive instances. They should try to falsify their hypotheses by seeking counterexamples. But the problem of scientific testing is really much deeper than simply the difficulty of defining a positive instance.

Consider the general hypothesis that "all emeralds are green". Surely a green emerald is a positive instance of this hypothesis. Now define the term "grue" as "green at time **t** and **t** is before 2100 AD or blue at **t** and **t** is after 2100 AD". Thus, after 2100 AD a cloudless sky will be grue, and any emerald already observed is grue as well. Consider the hypothesis "All emeralds are grue." It will turn out to be the case that every positive instance so far observed in favor of "All emeralds are green" is apparently a positive instance of "All emeralds are grue", even though the two hypotheses are incompatible in their claims about emeralds discovered after 2100 AD. But the conclusion that both hypotheses are equally well confirmed is absurd. The hypothesis "All emeralds are grue" is not just less well confirmed than "All emeralds are green", it is totally without evidential support altogether. But this means that all the green emeralds thus far discovered are not after all "positive instances" of "All emeralds are grue" – or else it would be a well-supported hypothesis since there are very many green emeralds and no non-green ones. But if green emeralds are not positive instances of the grue-hypothesis, then we need to give a reason why they are not.

We could restate the problem as one about falsification too. Since every attempt to falsify "All emeralds are green" has failed, it has also failed to falsify "All emeralds are grue." Both hypotheses have withstood the same battery of scientific tests. They are equally reasonable hypotheses. But this is absurd. The grue hypothesis is not one we would bother with for a moment, whether our method was seeking to confirm or to falsify hypotheses. So, our problem is not one that demanding science seek only falsification will solve.

One is inclined to respond to this problem by rejecting the predicate "grue" as an artificial, gerrymandered term that names no real property. "Grue" is constructed out of the "real properties" green and blue, and a scientific hypothesis must employ only real properties of things. Therefore, the grue-hypothesis is not a real scientific hypothesis and it has no positive instances. Unfortunately, this argument is subject to a powerful reply. Define bleen as "blue at **t** and **t** is earlier than 2100 AD or green at **t** when **t** is later than 2100 AD". We may now express the hypothesis that all emeralds are green as "all emeralds are grue at **t** and **t** is earlier than 2100 AD or bleen at **t** and **t** is later than 2100 AD". Thus, from the point of view of scientific language, "grue" is an intelligible notion. Moreover, consider the definition of "green" as "grue at **t** and **t** is earlier than 2100 AD or bleen at **t** and **t** is later than 2100 AD". What is it that prevents us from saying that green is the artificial, derived term, gerrymandered from "grue" and "bleen"?

What we seek is a difference between "green" and "grue" that makes "green" admissible in scientific laws and "grue" inadmissible. Following Nelson Goodman, who constructed the problem of "grue", philosophers have coined the term **"projectable"** for those predicates which are admissible in scientific laws. So, what makes "green" projectable? It cannot be that "green" is projectable because "All emeralds are green" is a well-supported law. For our problem is to show why "All emeralds are grue" is not a well-supported law, even though it has the same number of positive instances as "All emeralds are green." The puzzle of "grue", known as "the new riddle of induction", remains an unsolved problem in the theory of confirmation. Over the decades since its invention philosophers have offered many solutions to the problem, no one of which has gained ascendancy. But the inquiry has resulted in a far greater understanding of the dimensions of scientific confirmation than the logical positivists or their empiricist predecessors recognized. One thing all philosophers of science agree is that the new riddle shows how complicated the notion of confirmation turns out to be, even in the simple cases of generalizations about things we can observe.

3 Statistics and probability to the rescue?

At this point some scientists will lose patience with the philosopher of science. Why not simply treat the puzzle of grue and bleen as a philosopher's invention, and get on with the serious but perhaps more soluble problem of defining the notion of empirical confirmation. Even given the fallibility of science, the impossibility of establishing the truth or falsity of laws once and for all, and the role which auxiliary hypotheses inevitably play in the testing of theories, we explain how observation, data collection and experiment test scientific theory by turning to statistical theory and the notion of **probability**. The scientist who has lost patience with the heavy weather which philosophers make of how data confirm hypotheses will also insist that this is a problem for statistics, not philosophy. Instead of worrying about problems like what a positive instance of a hypothesis could be, or why positive instances confirm hypotheses we actually entertain and not an infinitude of alternative possibilities we haven't even dreamed up, we should leave the nature of hypothesis-testing to departments of probability and statistics. This is advice philosophers have resolutely tried to follow. As we shall see, it merely raises more problems about the way experience guides the growth of knowledge in science.

To begin with there is the problem of whether the fact that some data raise the probability of a hypothesis makes that data positive evidence for it. This may sound like a question trivially easy to answer, but it isn't. Define $p(h, b)$ as the probability of hypothesis h, given auxiliary hypotheses b, and $p(h, e$ and $b)$ as the probability of h, given the auxiliary hypotheses b, and some experimental observations e. Suppose we adopt the principle that

> e is positive evidence for hypothesis h if and only if $p(h, e$ and $b) > p(h, b)$

So, in this case, e is "new" data that count as evidence for h if they raise the probability of h (given the auxiliary assumptions required to test h). For example, the probability that the butler did it, h, given that the gun found at the body was not his, b, and the new evidence that the gun carried his fingerprints, is higher than the hypothesis that the butler did it, given the gun found at the body, and no evidence about fingerprints. It is the fingerprints that raise the probability of h. That's why the prints are "positive evidence".

It is easy to construct counterexamples to this definition of positive evidence which show that increasing probability is by itself neither necessary nor sufficient for some statement about observations to confirm a hypothesis. Here are two:

This book's publication increases the probability that it will be turned into a blockbuster film starring Emma Thompson. After all, were it never to have been published, the chances of its being made into a film would be even smaller than they are. But surely the actual publication of this book is not positive evidence for the hypothesis that this book will be turned into a blockbuster film starring Emma Thompson. It is certainly not clear that some fact which just raises the probability of a hypothesis thereby constitutes positive evidence for it. A similar conclusion can be derived from the following counterexample, which invokes lotteries, a useful notion when exploring issues about probability. Consider a fair lottery with 1000 tickets, 10 of which are purchased by Andy and 1 is purchased by Betty. h is the hypothesis that Betty wins the lottery. e is the observation that all tickets except those of Andy and Betty are destroyed before the drawing. e certainly increases the probability of h from .001 to .1. But it is not clear that e is positive evidence that h is true. In fact, it seems more reasonable to say that e is positive evidence that h is untrue, that Andy will win. For the probability that he wins has gone from .01 to .9. Another lottery case suggests that raising probability is not necessary for being positive evidence; indeed a piece of positive evidence may lower the probability of the hypothesis it confirms. Suppose in our lottery Andy has purchased 999 tickets out of 1000 sold on Monday. Suppose **e** is the evidence that by Tuesday 1001 tickets have been sold of which Andy purchased 999. This **e** lowers the probability that Andy will win the lottery from .999 to .998 ... But surely **e** is still evidence that Andy will win after all.

One way to deal with these two counterexamples is simply to require that **e** is positive evidence for **h** if **e** makes **h**'s probability high, say above .5. Then in the first case, since the evidence doesn't raise the probability of Betty's winning anywhere near .5, and in the second case the evidence does not lower the probability of Andy's winning much below .999, these cases don't undermine the definition of positive evidence when so revised. But of course, it is easy to construct a counterexample to this new definition of positive evidence as evidence that makes the hypothesis highly probable. Here is a famous case: **h** is the hypotheses that Andy is not pregnant, while **e** is the statement that Andy eats Weetabix breakfast cereal. Since the proba-bility of **h** is extremely high, **p(h, e)** – the probability of **h**, given **e** – is also extremely high. Yet **e** is certainly no evidence for **h**. Of course we have neglected the background information, **b**, built into the definition. Surely if we add the background information that no man has ever become pregnant, then **p(h, e and b)** – the probability of **h**, given **e** and **b** – will be the same as **p(h, e)**, and thus dispose of the counterexample. But if **b** is the statement that no man has ever become pregnant, and **e** is the statement that Andy ate Weetabix, and **h** is the statement that Andy is not pregnant, then **p(h, e and b)** will be very high, indeed about as close to 1 as a probability can get. So, even though **e** is not by itself positive evidence for **h**, **e** plus **b** is, just because **b** is positive evidence for **h**. We cannot exclude **e** as positive evidence, when **e** plus **b** is evidence, just because it is a conjunct which by itself has no impact on the probability of **h**, because sometimes positive evidence only does raise the probability of a hypothesis when it is combined with other data. Of course we want to say that in this case, **e** could be eliminated without reducing the probability of **h**, **e** is probabilistically irrelevant and that's why it is not positive evidence. But providing a litmus test for probabilistic irrele-vance is no easy task. It may be as difficult as defining a positive instance. In any case, we have an introduction here to the difficulties of expounding the notion of evidence in terms of the concept of probability.

Philosophers of science who insist that probability theory and its inter-pretation suffice to enable us to understand how data test hypotheses will respond to these problems that they reflect the mis-fit between proba-bility and our common-sense notions of evidence. Our ordinary concepts are qualitative, imprecise and not the result of a careful study of their implications. Probability is a quantitative mathematical notion with secure logical foundations. That enables us to make distinctions ordinary notions cannot draw, and to explain these distinctions. Recall the logical empiricists who sought rational reconstructions or explications of concepts like explanation that provide necessary and sufficient conditions in place of the imprecision and vagueness of ordinary language. Likewise, many contemporary students of the problem of confirmation seek a more precise substitute for the ordinary notion of evidence in the quantifiable notion of probability; for them, counterexamples such as the ones adduced above simply reflect the fact that the two concepts are not identical. They are no reason not to substitute "probability" for "evidence" in our inquiry about how data test theory. Some of these philosophers go further and argue that there is no such thing as evidence confirming or disconfirming a hypothesis by itself. Hypothesis testing in science is always a comparative affair: it only makes sense to say hypothesis h_1 is more or less well confirmed by the evidence than is hypothesis h_2, not that h_1 is confirmed by e in any absolute sense.

These philosophers hold that the mathematical theory of probability holds the key to understanding the confirmation of scientific theory. And this theory is extremely simple. It embodies only three very obvious assumptions:

1 Probabilities are measured in numbers from 0 to 1.
2 The probability of a necessary truth (like "4 is an even number") is 1.
3 If hypothesis **h** and **j** are incompatible (e.g. "There are 46 chromosomes in the human body" v. "there are 48 chromosomes in the human body), then $p(h \text{ or } j) = p(h) + p(j)$.

From these simple and straightforward assumptions the rest of the mathematical theory of probability can be derived by logical deduction alone. In particular, from these three axioms of the theory of probability, we can derive a theorem first proved by a British mathematician in the eighteenth century, Thomas Bayes, which has bulked large in contemporary discussions of confirmation. Before introducing this theorem, we need to define one more notion: conditional probability of any one statement, assuming the truth of another statement. The conditional probability of a hypothesis **h**, on a description of data **e**, written $p(h/e)$ is defined as the ratio of the probability of the truth of both **h** and **e** to the probability of the truth of **e** alone:

$$p(h/e) = df \; \frac{p(h \text{ and } e)}{p(e)}$$

Roughly "the conditional probability of **h** on **e**" measures the proportion of the probability that **e** is true which "contains" the probability that **h** is also true. Following Curd and Cover, we may explain this formula by imagining a dart-board with two overlapping circles on it, labeled **e** and **h**. If a dart thrown at the board lands in circle **e**, what is the probability that it has also landed inside circle **h**, i.e. the conditional probability that it has landed in the **h**-circle, assuming it landed in the **e**-circle? It will be the area which overlaps the **e**-circle and the **h**-circle, divided by the area of the **e**-circle. The larger the size of the **e**-circle compared to the size of the **h**-circle, the smaller the chance that by landing in the **e**-circle it also landed in the **h**-circle and vice versa. The conditional probability of **h** on **e**, i.e. the probability that **h** is true, given **e** is true, will be a fraction, whose numerator is the probability of **e** and whose denominator is the probability of **e** and **h** together. Notice that **p(h/e)** is not the same as **p(e/h)**.

Now if **h** is a hypothesis **h**, and **e** is a report of data, then we can calculate the conditional probability of **h** on **e**, **p(h/e)**. In other words, Bayes' theorem seems to give us just what we want: a mathematical formula for calculating how much more or less probable a bit of evidence **e**, makes any hypothesis **h**. The formula is as follows:

Bayes' theorem:

$$p(h/e) = \frac{p(e/h) \times p(h)}{p(e)}$$

Bayes' theorem tells us that once we acquire some data **e**, we can calculate how the data **e** change the probability of **h**, raising it or lowering, provided we already have three other numbers:

> **p(e/h)** – the probability that **e** is true assuming that **h** is true (not to be confused with **p(h/e)**, the probability that **h** is true, given **e**, which is what we are calculating). This number reflects the degree to which our hypothesis leads us to expect the data we have gathered. If the data are just what the hypothesis predicts then of course **p(e/h)** is very high. If the data are nothing like what the hypothesis predicts **p(e/h)** is low.

> **p(h)** – the probability of the hypothesis independent of the test to which the data described by **e** provides. If **e** reports new experimental data, then **p(h)** is just the probability the scientist assigned to **h** before the experiment was conducted.

p(e) – the probability that the statement describing the data is true independent of whether **h** is true or not. Where e is a surprising result which previous scientific theory and evidence (independent of **h**) does not lead us to expect, **p(e)** will be low.

Two simple examples may help us see how Bayes' theorem is supposed to work: Consider how data on the observed position of Halley's comet provide a test for Newton's laws. Suppose, given prior observations, that **p(e)**, the probability that Halley's comet will be observed in particular location of the night sky, is .8. This allows for imperfections in the telescope, atmospheric irregularities, all the factors that eventually led astronomers to take many photographs of the stars and planets and to average their positions to make estimates of their expected positions in the sky. **p(e/h)** is also high, the expected position of Halley's comet in the night sky is very close to what the theory predicts it would be. Let's set **p(e/h)** at .95. Let's assume that prior to the acquisition of **e**, the new data about Halley's comet, the probability that Newton's laws are true is, say, .8. Thus, if Halley's comet appears where expected, **p(h/e)** = (.95 ...) × (.8)/ (.8) = .95. Thus, the evidence as described by **e** has raised the probability of Newton's laws from .8 to .95.

But now, suppose we acquire new data about, say, the precession of the perihelion of mercury – that is, data which show that the elliptical orbit of Mercury around the sun is itself swinging so that the closest point between Mercury and the sun keeps shifting. Suppose, as was indeed the case, that the figure turns out to be much higher than Newton's laws (and the auxiliary hypotheses used to apply them) would lead us to expect, so that **p(e/h)** is low, say .3. Since Newton's laws did not lead us to expect these data, the **prior probability** of **e** must be low, so let's let **p(e)** be low, say, .2; and the prior probability of such unexpected data, given Newton's laws plus auxiliary hypotheses, will also be quite low, say, **p(e/h)** is .1. If **p(h)** for Newton's laws plus auxiliaries is .95, then Bayes' theorem tells us that for the new **e**, the precession data for Mercury, the **p(h/e)** = (.1) × (.95) / (.2) = .475, a significant drop from .95. Naturally, recalling the earlier success of Newton's laws in uncovering the existence of Neptune and Uranus, the initial blame for the drop was placed on the auxiliary hypotheses. Bayes' theorem can even show us why. Though the numbers in our example are made up, in this case, the auxiliary assumptions were eventually vindicated, and the data about the much greater than expected precession of the perihelion of Mercury undermined Newton's theory, and (as another application of Bayes' theorem would show), increased the probability of Einstein's alternative theory of relativity.

Philosophers and many statisticians hold that the reasoning scientists use to test their hypotheses can be reconstructed as inferences in accordance with Bayes' theorem. These theorists are called Bayesians. Some philosophers and historians of science among them seek to show that the history of acceptance and rejection of theories in science honors Bayes' theorem, thus showing that, in fact, theory-testing has been on a firm footing all along. Other philosophers, and statistical theorists, attempt to apply Bayes' theorem actually to determine the probability of scientific hypotheses when the data are hard to get, sometimes unreliable, or only indirectly relevant to the hypothesis under test. For example, they seek to determine the probabilities of various hypotheses about evolutionary events like the splitting of ancestral species from one another, by applying Bayes' theorem to data about differences in the polynucleotide sequences of the genes of currently living species.

How much understanding of the nature of empirical testing does **Bayesianism** really provide? Will it reconcile science's empiricist epistemology with its commitment to unobservable events and processes that explain observable ones. Will it solve Hume's problem of induction? To answer these questions, we must first understand what the probabilities are that all these **p**s symbolize and where they come from. We need to make sense of **p(h)**, the probability that a certain proposition is true. There are at least two questions to be answered. First, there is the "metaphysical" question of what fact it is about the world, if any, that makes a particular probability value **p(h)**, for a hypothesis **h**, the true or correct one? Second, there is the epistemological question of justifying our estimate of this probability value. The first question may also be understood as a question about the meaning of probability statements, and the second about how they justify inductive conclusions about general theories and future eventualities.

Long before the advent of Bayesianism in the philosophy of science, the meaning of probability statements was already a vexed question. There are some traditional interpretations of probability we can exclude as unsuitable interpretations for the employment of Bayes' theorem. One such is the interpretation of probability as it is supposed to figure in fair games of chance like roulette or black jack. In a fair game of roulette the chance of the ball landing in any trap is exactly 1/37 or 1/38 because there are 37 (or in Europe 38) traps into which the ball can land. Assuming it is a fair roulette wheel, the probability of the hypothesis that the ball will land on number 8 is exactly 1/37 or 1/38 and we know this *a priori* – without experience – because we know *a priori* how many possibilities there are and that each is equally probable (again, assuming the roulette wheel is fair, a bit of knowledge we could never have acquired *a priori* anyway!). Now, when it comes to hypotheses that can account for a finite body of data, there is no limit to the number of possibilities and no reason to think that each of them has the same probability. Accordingly, the probabilities of a hypothesis about, say, the number of chromosomes in a human nucleus, will not be determinable *a priori*, by counting up possibilities and dividing 1 by the number of possibilities.

Another interpretation of probabilities involves empirical observations, for example, coin flips. To establish the frequency with which a coin will come up heads, one flips it several times and divides the number of times it comes up heads by the number of times it was flipped. When will this frequency be a good estimate of the probability of heads? When the number of coin flips is large, and the frequencies we calculate for finite numbers of coin flips converge on one value and remain near that value no matter how many times we continue flipping. We can call this value, if there is one, the **long-run relative frequency** of heads. And we treat it as a measure of the probability the coin comes up heads. But is the long-run relative frequency of heads identical to the probability it will come up heads? This sounds like a silly question, until you ask what the connection is between the long-run relative frequency's being, say, 50 per cent and the chance that the very next toss will be heads. Notice that a long-run relative frequency of 50 per cent is compatible with a run of ten, or a hundred, or a million heads in a row, just so long as the total number of tosses is very large, so large that a million is a small number in comparison to the total number of tosses. If this is right, the long-run relative frequency is compatible with any finite run of all heads, or all tails, and of course perfectly compatible with the coin's coming up tails on the next toss. Now, suppose we want to know what the probability is that the coin will come up heads on the next toss. If the probability that the coin will come up heads on the next toss is a property of that particular toss, it is a different thing from the long-run relative frequency of heads (which is perfectly compatible with the next 234,382 tosses all being tails). We need some principle that connects the long run to the next toss. One such principle which gets us from the long-run relative frequency to the probability of the next toss being heads is to assume that coins do in any finite run what they do in the long run. But this principle is just false. A better principle for connecting long-run relative frequencies to the probability of the next occurrence is something like this: If you know the long-run relative frequency then you know how to bet on whether the coin will land heads or tails, and if you take all bets against heads at odds greater than even money, you will win. But notice this is a conclusion about what you should do as a gambler, not a conclusion about what the coin will in fact do. We will come back to this insight.

Could long-run relative frequencies provide the probability values for a hypothesis without a track record? It is hard to see how. Compare a novel hypothesis to a shiny new penny about to be flipped. Long-run relative frequencies data provide some reason to ascribe a probability of

50 per cent to the chances of heads on the new penny. Is there a track record of previous hypotheses relevant to the new one? Only if we can compare it to the right class of similar hypotheses the way we can compare new pennies to old ones. But hypotheses are not like pennies. Unlike pennies, they differ from one another in ways we cannot quantify as we would have to were we to grade them for similarity to one another. Even if we could identify the track record of truth and falsity for similar hypotheses formulated over the past history of science, we would have the problems of (a) justifying the inference from a finite actual sequence to a long-run relative frequency, and (b) justifying the inference from a long-run relative frequency to the next case, the new hypothesis. Recall, that in the case of coin-flipping, the only connection appears to be that relative frequencies are our best guide to how to lay our bets about the next toss. Perhaps the kind of probability which theory-testing invokes is the gambler's kind, what has come to be called "subjective probability". "Subjective" because it reflects facts about the gambler, and what the gambler believes about the past and the future, and "probability" because the bets the gambler makes should honor the axioms of probability.

It is the claim that, in scientific testing, the relevant probabilities are subjective probabilities – that is, gambler's odds – that is the distinctive mark of the Bayesian. A Bayesian is someone who holds that at least two of the three probabilities we need to calculate $p(h/e)$ are just a matter of betting odds and that within certain weak constraints, they can take on any values at all. You and I may think that the best betting odds are those which mirror our previous experience of actual frequencies or our estimate of long-run relative frequencies, but this is no part of Bayesianism. The Bayesian holds that in the long run it doesn't matter what values they start with, Bayes' theorem will lead the scientist inexorably to the (available) hypothesis best supported by the evidence. These remarkable claims demand explanation and justification.

Calculating the value of $p(e/h)$ is a matter of giving a number to the probability that e obtains if h is true. This is usually easy to do. If h tells us to expect e, or data close to e, then $p(e/h)$ will be very high. The problem is that using Bayes' theorem also requires we calculate input values, so-called "prior probabilities", $p(h)$ and $p(e)$. $p(h)$ is especially problematical: after all, if h is a new theory no one has ever thought of, why should there be any particular right answer to the question of with what probability it is true. And assigning a value to $p(e)$, the probability that our data description is correct, may involve so many auxiliary assumptions that even if there is a correct number it is hard to see how we could figure out what it is. The Bayesian asserts that these are not problems. Both values, $p(h)$ and $p(e)$ (and $p(e/h)$ for that matter), are simply degrees of belief, and degrees of belief are simply a matter of what betting odds the scientist would take or decline on whether their beliefs are correct. The higher the odds one takes, the stronger the degree of belief. Here the Bayesian takes a page from economists and others who developed the theory of rational choice under uncertainty. The way to measure a degree of belief is to offer the believer wagers against the truth of his or her belief. Other things being equal, if you are rational, and you are willing to take a bet that **h** is true at odds of 4:1 against you, then your degree of belief that **h** is true is .8. If you are willing to take a bet at 5:1 against you, then your degree of belief is .833. Probabilities are identical to degrees of belief. The other things that have to be equal for this way of measuring the strength of your beliefs are (a) that you have enough money, so that you are not so averse to the risk of losing that it swamps your attraction to the prospect of winning, (b) that the degrees of belief you assign to your beliefs obey the rules of logic and the three laws of probability above. So long as your degrees of belief, a.k.a. probability assignments, honor these two assumptions, the Bayesian says, the initial values or "prior probabilities" you assign to them can be perfectly arbitrary, in fact may be arbitrary, but it doesn't really matter. In the parlance of the Bayesians, as more and more data come in, the prior probabilities will be "swamped", that is, when we use Bayes' theorem to "update" prior probabilities, i.e. feed new **p(e)**s into the latest values for **p(e/h)** and **p(e/h)**, the successive values of **p(h/e)** will converge on the correct value, no matter what initial values for these three variables we start with! Prior probabilities are nothing but measures of the individual scientist's purely subjective degree of belief before applying Bayes' theorem. In answer to our metaphysical question about what facts about the world probabilities report, prior probabilities report no facts about the world, or at least none about the world independent of our beliefs. In answer to the epistemological question of what justifies our estimates of probabilities, when it comes to prior probabilities, no more justification is needed or possible than that our estimates obey the axioms of probability.

There is no right answer or wrong answer as to what the prior probabilities of **p(h)** or **p(e)** are, so long as the values of these probabilities obey the rules of probability and logical consistency on betting. Logical consistency simply means that one places one's bets – that is, assigns strengths to one's degrees of belief – in such a way that bookies can't use you for a money pump: that is, make bets with you so that no matter which propositions come out true or false you lose money. What is more, another theorem of the probability theory shows that if we apply Bayes' theorem relentlessly to "update" our prior probabilities as new evidence comes in, the value of **p(h)** all scientists assign will converge on a single value no matter where each scientist begins in his or her original assignment of prior probabilities. So not only are prior probabilities arbitrary but it doesn't matter that they are! Some scientists may assign prior probabilities on considerations like simplicity or economy of assumptions, or similarity to already proven hypotheses, or symmetry of the equations expressing the hypothesis. Other scientists will assign prior probabilities on the basis of superstition, aesthetic preference, number worship, or by pulling a ticket out of a hat. It doesn't matter, so long as they all conditionalize on new evidence via Bayes' theorem.

It is not much of an objection to this account of scientific testing that scientists actually offer good reasons for their methods of assigning of prior probabilities. To begin with, Bayesianism doesn't condemn these reasons, at worst it is silent on them. But if features like the simplicity of a hypothesis or the symmetry of its form do in fact increase its prior probability, this will be because a hypothesis having features like this will, via Bayes' theorem, acquire a higher posterior probability than other hypotheses with which it is competing that lack these features. More important, attempts to underwrite the reasoning of scientists who appeal to considerations like economy, simplicity, symmetry, invariance or other formal features of hypotheses, by claim that such features increase the objective probability of a hypothesis, come up against the problem that the only kind of probability that seems to make any sense for scientific testing is Bayesian subjective probability.

Furthermore, so understood, some Bayesians hold that probabilities can after all deal with some of the traditional problems of confirmation. Recall the black boot/white swan positive-instance puzzle discussed above, according to which a black boot is positive evidence for "All swans are white." Not on Bayesianism. After all, the prior conditional probability of a boot being black, conditional on all swans being white, is lower than the prior probability of the next swan we see being white, conditional on all swans being white, which is high. When we plug these two priors into Bayes' theorem, if the prior probabilities of seeing a white swan and a black boot are equal, the probability of "all swans are white" is raised much more by the latter conditional probability.

One of the major problems confronting Bayesianism, and perhaps other accounts of how evidence confirms theory, is the "problem of old evidence". It is not uncommon in science for a theory to be strongly confirmed by data already well known long before the hypothesis was formulated. Indeed, as we will see in Chapter 6, this is an important feature of situations in which scientific revolutions take place: Newton's theory was strongly confirmed by its ability to explain the data on which Galileo's and Kepler's theories were based. Einstein's general theory of relativity explained previously recognized but highly unexpected data such as the invariance of the speed of light and the precession of the perihelion of Mercury. In these two cases $p(e) = 1$, $p(e/h)$ is very high.

Plugging these values into Bayes' theorem gives us

$$p(h/e) = \frac{1 \times p(h)}{1} = p(h)$$

In other words, on Bayes' theorem the old evidence does not raise the posterior probability of the hypothesis – in this case Newton's laws, or the special theory of relativity – at all. Bayesians have gone to great lengths to deal with this problem. One stratagem is to "bite the bullet" and argue that old evidence does not in fact confirm a new hypothesis. This approach makes common cause with the well-established objection to hypotheses which are designed with an eye to available evidence. Scientists who construct hypotheses by intentional "curve-fitting" are rightly criticized and their hypotheses are often denied explanatory power on the grounds that they are *ad hoc*. The trouble with this strategy is that it doesn't so much solve the original Bayesian problem of old evidence as combine it with another problem: how to distinguish cases like the confirmation of Newton's and Einstein's theories by old evidence from cases in which old evidence does not confirm a hypothesis because it was accommodated to the old evidence. The alternative approach to the problem of old evidence is to supplement Bayes' theorem with some rule that gives **p(e)** a value different from 1. For example, one might try to give **p(e)** the value it might have had before **e** was actually observed in the past, or else try to rearrange one's present scientific beliefs by deleting **e** from them and anything which **e** makes probable; then go back and assign a value to **p(e)**, which presumably will be lower than 1. This strategy is obviously an extremely difficult one to actually adopt. And it is (subjectively) improbable that any scientist consciously thinks this way.

Many philosophers and scientists who oppose Bayesianism do so not because of the difficulties which are faced by the program of developing it as an account of the actual character of scientific testing. Their problem is with the approach's commitment to subjectivism. The Bayesian claim that no matter what prior probabilities the scientist subjectively assigns to hypotheses, their subjective probabilities will converge on a single value, is not sufficient consolation to opponents. Just for starters, values of **p(h)** will not converge unless we start with a complete set of hypotheses that are exhaustive and exclusive competitors. This seems never to be the case in science. Moreover, objectors argue, there is no reason given why the value on which all scientists will converge by Bayesian conditionalization is the *right* value for **p(h)**. This objection of course assumes there is such a thing as the right, i.e. the objectively correct, probability, and so begs the question against the Bayesian. But it does show that Bayesianism is no solution to Hume's problem of induction, as a few philosophers hoped it might be.

And the same pretty much goes for other interpretations of probability. If sequences of events reveal long-run relative frequencies that converge on some probability value and stay near it forever, then we could rely on them at least for betting odds. But to say that long-run relative frequencies will converge on some value is simply to assert that nature is uniform, that the future will be like that past, and so begs Hume's question. Similarly, hypothesizing probabilistic propensities that operate uniformly across time and space also begs the question against Hume's argument. In general, probabilities are useful only if induction is justified, not vice versa. Still, as noted, only a handful of philosophers have sought explicitly to solve Hume's problem by appeal to probabilities.

There is a more severe problem facing Bayesianism. It is the same problem that we came up against in the discussion of how to reconcile empiricism and explanation in theoretical science. Because empiricism is the doctrine that knowledge is justified by observation, in general, it must attach the highest probability to statements which describe observations, and lower probability to those which make claims about theoretical entities. Since theories explain observations, we may express the relation between theory and observation as (t and $t \to h$) – where t is the theory and $t \to h$ reflects the explanatory relation between the theoretical claims of the theory t and an observational generalization h, describing the data that the theory leads us to expect. The relation between t and h may be logically deductive, or it may be some more complex relation. But $p(h)$ must never be lower than $p(t$ and $t \to h)$, just because the antecedent of the latter is a statement about what cannot be observed whose only consequence for observation is h. Bayesian conditionalization on evidence will never lead us to prefer (t and $t \to h$) to h alone. But this is to say that Bayesianism cannot account for why scientists embrace theories at all, instead of just according high subjective probability to the observational generalizations that follow from them. Of course, if the explanatory power of a theory were a reason for according it a high prior probability, then scientists' embracing theories would be rational from the Bayesian point of view. But to accord explanatory power such a role in strengthening the degree of belief requires an account of explanation. And not just any account. It cannot for example make do with the D-N model, for the principal virtue of this account of explanation is that it shows that the explanandum phenomenon could be expected with at least high probability. In other words, it grounds explanatory power on strengthening probability, and so cannot serve as an alternative to probability as a source of confidence in our theories. To argue, as seems tempting, that our theories are explanatory in large part because they go beyond and beneath observations to their underlying mechanisms, is something the Bayesian cannot do.

4 Underdetermination

The testing of claims about unobservable things, states, events or processes is evidently a complicated affair. In fact the more one considers how observations confirm hypotheses and how complicated the matter is the more one is struck by a certain inevitable and quite disturbing "underdetermination" of theory by observation.

As we have noted repeatedly, the "official epistemology" of modern science is empiricism, the doctrine that our knowledge is justified by experience – observation, data collection, experiment. The objectivity of science is held to rest on the role which experience plays in choosing between hypotheses. But if the simplest hypothesis comes face to face with experience only in combination with other hypotheses, then a negative test may be the fault of one of the accompanying assumptions; a positive test may reflect compensating mistakes in two or more of the hypotheses involved in the test that cancel one another out. Moreover, if two or more hypotheses are always required in any scientific test, then when a test-prediction is falsified there will always be two or more ways to "correct" the hypotheses under test. When the hypothesis under test is not a single statement like "all swans are white" but a system of highly theoretical claims like the kinetic theory of gases, it is open to the theorist to make one or more of a large number of changes in the theory in light of a falsifying test, any one of which will reconcile the theory with the data. But the large number of changes possible introduces a degree of arbitrariness foreign to our picture of science. Start with a hypothesis constituting a theory that describes the behavior of unobservable entities and their properties. Such a hypothesis can be reconciled with falsifying experience by making changes in it that cannot themselves be tested except through the same process all over again – one which allows for a large number of further changes in case of falsification. It thus becomes impossible to establish the correctness or even the reasonableness of one change over another. Two scientists beginning with the same theory, subjecting it to the same initial disconfirming test, and repeatedly "improving" their theories in the light of the same set of further tests, will almost certainly end up with completely different theories, both equally consistent with the data their tests have generated.

Imagine, now, the "end of inquiry" when all the data on every subject are in. Can there still be two distinct equally simple, elegant, and otherwise satisfying theories equally compatible with all the data, and incompatible with one another? Given the empirical slack present even when all the evidence appears to be in, the answer seems to be that such a possibility cannot be ruled out. Since they are distinct theories, our two total "systems of the world" must be incompatible, and therefore cannot both be true. We cannot remain either agnostic about whether one is right or ecumenical about embracing both. Yet it appears that observation would not be able to decide between these theories.

In short, theory is underdetermined by observation. And yet science does not show the sort of proliferation of theory and the kind of unresolvable theoretical disputes that the possibility of this underdetermination might lead us to expect. But the more we consider reasons why this sort of underdetermination does not manifest itself, the more problematical becomes the notion that scientific theory is justified by objective methods that make experience the final court of appeal in the certification of knowledge. For what else besides the test of observation and experiment could account for the theoretical consensus characteristic of most natural sciences? Of course there are disagreements among theorists, sometimes very great ones, and yet over time these disagreements are settled, to almost universal satisfaction. If, owing to the ever-present possibility of underdetermination, this theoretical consensus is not achieved through the "official" methods, how is it achieved?

Well, besides the test of observation, theories are also judged on other criteria: simplicity, economy, consistency with other already adopted theories. But these criteria simply invoke observations – albeit somewhat indirectly. A theory's consistency with other already well-established theories confirms that theory only because observations have established the theories it is judged consistent with. Simplicity and economy in theories are themselves properties that we have observed nature to reflect and other well-confirmed theories to bear, and we are prepared to surrender them if and when they come into conflict with our observations and experiments. One alternative source of consensus philosophers of science are disinclined to accept is the notion that theoretical developments are epistemically guided by non-experimental, non-observational considerations, such as *a priori* philosophical commitments, religious doctrines, political ideologies, aesthetic tastes, psychological dispositions, social forces or intellectual fashions. Such factors we know will make for consensus, but not necessarily one that reflects increasing approximation to the truth, or to objective knowledge. Indeed, these non-epistemic, non-scientific forces and factors are supposed to deform understanding and lead away from truth and knowledge.

The fact remains that a steady commitment to empiricism coupled with a fair degree of consensus about the indispensability of scientific theorizing strongly suggests the possibility of a great deal of slack between theory and observation. But the apparent absence of arbitrariness fostered by underdetermination demands explanation. And if we are to retain our commitment to science's status as knowledge *par excellence*, this explanation had better be one we can parlay into a justification of science's objectivity as well. The next chapter shows that prospects for such an outcome are clouded with doubt.

Summary

Empiricism is the epistemology which has tried to make sense of the role of observation in the certification of scientific knowledge. Since the eighteenth century, if not before, especially British philosophers like Hobbes, Locke, Berkeley and Hume have found inspiration in science's successes for their philosophies, and sought philosophical arguments to ground science's claim. In so doing, these philosophers and their successors set the agenda of the philosophy of science and revealed how complex is the apparently simple and straightforward relation between theory and evidence.

In the twentieth century the successors of the British empiricists, the logical positivists or "logical empiricists" as some of them preferred, sought to combine the empiricist epistemology of their predecessors with advances in logic, probability theory and statistical inference, to complete the project initiated by Locke, Berkeley and Hume. What they found was that some of the problems seventeenth- and eighteenth-century empiricism uncovered were even more resistant to solution when formulated in updated logical and methodological terms. "Confirmation theory", as this part of the philosophy of science came to be called, has greatly increased our understanding of the "logic" of confirmation, but has left as yet unsolved Hume's problem of induction, the further problem of when evidence provides a positive instance of a hypothesis, and the "new riddle of induction" – Goodman's puzzle of "grue" and "bleen."

Positivists and their successors have made the foundations of probability theory central to their conception of scientific testing. Obviously much formal hypothesis testing employs probability theory. One attractive late twentieth-century account that reflects this practice is known as Bayesianism. This view holds that scientific reasoning from evidence to theory proceeds in accordance with Bayes' theorem about conditional probabilities, under a distinctive interpretation of the probabilities it employs.

The Bayesians hold that scientists' probabilities are subjective degrees of belief or acceptance of a claim – betting odds. By contrast with other interpretations, according to which probabilities are long-run relative frequencies, or distributions of actualities among all logical possibilities, this frankly psychological interpretation of probability is said to best fit the facts of scientific practice and its history.

The Bayesian responds to complaints about the subjective and arbitrary nature of the probability assignment it tolerates by arguing that, no matter where initial probability estimates start out, in the long run using Bayes' theorem on all possible alternative hypotheses will result in their convergence on the most reasonable probability values, if there are such values. Bayesianism's opponents demand that it substantiate the existence of such "most reasonable" values and show that all alternative hypotheses are being considered. To satisfy these demands would be tantamount to solving Hume's problem of induction. Finally, Bayesianism has no clear answer to the problem which drew our attention to hypothesis-testing: the apparent tension between science's need for theory and its reliance on observation.

This tension expresses itself most pointedly in the problem of underdetermination. Given the role of auxiliary hypotheses in any test of a theory, it follows that no single scientific claim meets experience for test by itself. It does so only in the company of other, perhaps large numbers of, other hypotheses needed to effect the derivation of some observational prediction to be checked against experience. But this means that a disconfirmation test, in which expectations are not fulfilled, cannot point the finger of falsity at one of these hypotheses and that adjustments in more than one may be equivalent in reconciling the whole package of hypotheses to observation. As the size of a theory grows, and it encompasses more and more disparate phenomena, the alternative adjustments possible to preserve or improve it in the face of recalcitrant data increase. Might it be possible, at the never-actually-to-be-reached "end of inquiry", when all the data are in, that there be two distinct total theories of the world, both equal in evidential support, simplicity, economy, symmetry, elegance, mathematical expression or any other desideratum of theory choice? A positive answer to the question may provide powerful support for an instrumentalist account of theories. For apparently there will be no fact of the matter accessible to inquiry that can choose between the two theories.

And yet, the odd thing is that underdetermination is a mere possibility. In point of fact, it almost never occurs. This suggests two alternatives. The first alternative, embraced by most philosophers of science, is that observation really does govern theory choice (or else there would be more competition among theories and models than there is); it's just that we simply haven't figured it all out yet. The second alternative is more radical, and is favored by a generation of historians, sociologists of science and a few philosophers who reject both the detailed teachings of logical empiricism and also its ambitions to underwrite the objectivity of science. On this alternative, observations underdetermine theory, but it is fixed by other facts – non-epistemic ones, like bias, faith, prejudice, the desire for fame, or at least security, and power politics. This radical view, that science is a process, like other social processes, and not a matter of objective progress, is the subject of the next two chapters.

Questions

1 Discuss critically: "Lots of scientists pursue science successfully without any regard to epistemology. The idea that science has an 'official one', and that empiricism is it, is wrong-headed."
2 Why would it be correct to call Locke the father of modern scientific realism and Berkeley the originator of instrumentalism? How would Berkeley respond to the argument for realism as an inference to the best explanation of science's success?
3 We have defined grue and bleen by way of the concepts of green and blue. Construct a definition of green and blue which starts with grue and bleen. What does this show about the projectability of green and blue?
4 Give examples, preferably from science, in which all three concepts of probability are used: subjective, relative frequency, and probabilistic propensity. Hint: think of weather reports.
5 Argue against the claim that two equally well confirmed total theories which appear to be incompatible are only disguised terminological variants of one another.

Further reading

The relationship between science and philosophy, and especially the role of science in the dispute between empiricism and rationalism during the early modern period, are treated in E. A. Burtt, *Metaphysical Foundations of Modern Science*. John Locke's *Essay on Human Understanding* is a long work; George Berkeley's *Principles of Human Knowledge* is brief but powerful. The last third develops an explicitly instrumental conception of science which he contrasts to Locke's realism. Berkeley argued for idealism – the thesis that only what is perceived exists, that the only thing we perceive is ideas, that therefore only ideas exist. About this work, Hume wrote "it admits no refutation, and carried no conviction" in his *Inquiry Concerning Human Understanding*. In this work he develops the theory of causation discussed in Chapter 2, the theory of language common to classical empiricists and logical positivists, and the problem of induction.

J. S. Mill, *A System of Logic*, carried the empiricist tradition forward in the nineteenth century, and proposed a canon for experimental science still widely employed under the name, Mill's methods of induction. The physicist Ernst Mach, *The Analysis of Sensation*, embraced Berkeley's attack on theory as empirically unfounded against Ludwig Boltzman's atomic theory. This work was greatly influential on Einstein. In the first half of the twentieth century, logical empiricists developed a series of important theories of confirmation – R. Carnap, *The Continuum of Inductive Methods*; H. Reichenbach, *Experience and Prediction*. Their younger colleagues and students wrestled with these theories and their problems. Essays on confirmation theory in Hempel, *Aspects of Scientific Explanation*, are of special importance, as is N. Goodman, *Fact, Fiction and Forecast*, where the new riddle of induction is introduced along with Goodman's path-breaking treatment of counterfactuals.

W. Salmon, *The Foundations of Scientific Inference*, is a useful introduction to the history of confirmation theory from Hume through the positivists and their successors. D. C. Stove, *Hume, Probability and Induction*, attempts to solve the problem of induction probabilistically.

Objection to the logical empiricist theory of testing was early advanced by Karl Popper, *Logic of Scientific Discovery*, first published in German in 1935. Carnap argued the method of falsification as the key to scientific objectivity. The arguments of W. V. O. Quine, *From a Logical Point of View*, and *Word and Object* – following a much earlier work, P. Duhem, *The Aim and Structure of Physical Theory*, that the role of auxiliary hypotheses makes strict falsification impossible – limited the influence of Popper's views.

L. Savage, *Foundations of Statistics*, provides a rigorous presentation of Bayesianism, as does R. Jeffrey, *The Logic of Decision*. A philosophically sophisticated presentation is P. Horwich, *Probability and Evidence*.

The problem of old evidence, among other issues, has led to dissent from Bayesianism by C. Glymour, *Theory and Evidence*.

Peter Achinstein, *The Concepts of Evidence*, anthologizes several papers that reflect the complexities of inference from evidence to theory.

The possibility of underdetermination is broached first in Quine, *Word and Object*. It has been subject to sustained critical scrutiny over the succeeding half-century. For an important example of this criticism, see J. Leplin and L. Laudan, "Empirical Equivalence and Underdetermination"; C. Hoefer and A. Rosenberg, "Empirical Equivalence, Underdetermination and Systems of the World", respond to their denial of underdetermination.

CHAPTER 6
The challenge of history and post-positivism

Overview

If observational evidence underdetermines theories, we need at least an explanation of what does determine the succession of theories which characterizes science's history. Even more, for philosophy's purposes, we need a justification for the claim that these observationally unsupported theories are epistemically rational and reasonable ones to adopt. Clearly, empiricism cannot by itself do this, as its resources in justification are limited to observation.

Thomas Kuhn, an important historian of science, was among the first to explore the history of science for these non-observational factors that explain theory-choice, and to consider how they might justify it as well. His book, *The Structure of Scientific Revolutions*, sought to explore the character of scientific change – how theories succeed one another – with a view to considering what explains and what justifies the replacement of one theory by another. The logical empiricists hold that theories succeed one another by reduction, which preserves what is correct in an earlier theory, and so illuminates the history of science as progress. Kuhn's research challenges this idea.

By introducing considerations from psychology and sociology as well as history, Kuhn reshaped the landscape in the philosophy of science and made it take seriously the idea that science is not a disinterested pursuit of the truth, successively cumulating in the direction of greater approximation to the truth, as guided by unambiguous observational test.

Kuhn's shocking conclusion suggests that science is as creative an undertaking as painting or music, and not to be viewed as more objectively progressive, correct or approximating to some truth about the world than these other human activities. The history of science is the history of change, but not progress; in a sense that Kuhn defends, we are no nearer the truth about the nature of things nowadays than we were in Aristotle's time. These shocking conclusions represent a great challenge to contemporary philosophy of science.

Much of the philosophical underpinnings for views like Kuhn's can be found in the work of an equally influential philosopher, W. V. O. Quine, who attacked logical empiricism "from within", so to speak. A student of the logical empiricists, Quine was among the first to see that the epistemology underlying their philosophy of science could not satisfy its own requirements for objective knowledge, and was based on a series of unsupportable distinctions. By casting doubt on the foundations of a tradition in philosophy that went back to Locke, Berkeley and Hume, Quine made it impossible for philosophers of science to ignore the controversial claims of Kuhn and those sociologists, psychologists and historians ready to employ his insights to uncover the status of science as a "sacred cow".

1 A place for history?

In the last chapter we traced the development of philosophy's traditional analysis of scientific knowledge as the outcome of attempts to
explain our observations which are themselves "controlled" by our
observations. Empiricism, the ruling "ideology" of science, assures us that
what makes scientific explanations credible, and what insures the self-
correction of science, as well as its ever-increasing predictive powers, is the
role that observation, experiment and test play in the certification of
scientific theory.

But we have also seen that actually making this role precise is not
something the philosophy of science has been able to do. Not only can
philosophy of science not provide an uncontroversial empiricist justification for our knowledge of the existence of theoretical entities, it cannot
even assure that the terms that name these entities are meaningful. Even
worse, the simplest evidential relation between a piece of data and a
hypothesis which that data might test seems equally difficult to express
with the sort of precision that both science and the philosophy of science
seem to require. One might hold that this is not a problem for scientists,
just for philosophers of science. After all, we know that theoretical terms
are indispensable, because theoretical entities exist and we need to invoke
them in explanations and predictions. And we know that scientific
hypotheses' abilities to withstand empirical test is what makes them
knowledge. Formalizing these facts may be an interesting exercise for
philosophy but it need not detain the working scientist.

This would be a superficial view of the matter. To begin with, it would
be a double standard not to demand the same level of detail and precision
in our understanding of science as science demands of itself in its understanding of the world. Scientific empiricism bids us test our ideas against
experience; we cannot do this if these ideas are vague and imprecise. The
same must go for our ideas about the nature of science itself. Second, if
we cannot provide a precise and detailed account of such obvious and
straightforward matters as the existence of theoretical entities and the
nature of scientific testing, then this is a symptom that there may be
something profoundly wrong in our understanding of science. This will
be of particular importance to the extent that less well-developed disciplines look to the philosophy of science for guidance, if not recipes on
how to be scientific.

The dissatisfaction with philosophy of science's answers to fundamental questions about theories and their testing of course led
philosophers of science to begin rethinking the most fundamental presuppositions of the theory of science embodied in logical empiricism. The
re-examination began with the uncontroversial claim that the philosophy

of science should provide a picture of the nature of science that mirrors what we know about its history and its actual character. This may sound uncontroversial until it is recalled how much traditional philosophy of science relied on considerations from formal logic coupled with a narrow range of examples from physics.

Among the earliest, and certainly the most influential document in the reconsideration of the nature of science from the perspective of its history, was Thomas Kuhn's *The Structure of Scientific Revolutions*. This slim work set out to bring the philosophy of science face to face with important episodes from its history. But it ended up completely undermining philosophy's confidence that it understood anything about science. And it became the single most heavily cited work in the second half of the twentieth century's absorption with science. How could this have happened?

The study of the history of science since well before Newton suggested to Kuhn that claims about the world we might now view as pre- or unscientific myths were embraced by learned people whose aim was to understand the world for much the same sort of reasons that we embrace contemporary physical theory. If it is the sort of reasons that support a belief which makes it scientific, then these myths were science too. Or alternatively, our latest scientific beliefs are myths, like the pre- and unscientific ones they replaced. Kuhn held that the first of these alternatives was to be preferred. Adopting this perspective makes the history of long-past science an important source of data in any attempt to uncover the methods that make science objective knowledge. The second alternative, that contemporary science is just the latest successor in a sequence of mythic "world-views", no more "objectively true" than its predecessors, seemed to most philosophers of science (if not always to Kuhn), preposterous. The trouble is that Kuhn's account of the nature of science was widely treated outside philosophy of science as having supported this second alternative at least as much as the first one.

Kuhn's ostensible topic was scientific change, how the broadest theories replace one another during periods of scientific revolution. Among the most important of these was the shift from Aristotelian physics to Newtonian mechanics, from phlogiston chemistry to Lavoisier's theories of reduction and oxidation, from non-evolutionary biology to Darwinism, and from Newtonian mechanics to relativistic and quantum mechanics. Periods of revolutionary change in science alternate with periods of what Kuhn called "**normal science**", during which the direction, the methods, the instruments and the problems that scientists face are all fixed by the established theory. But Kuhn considered that the term "theory" did not aptly describe the intellectual core of a program of "normal science". Instead he coined the term "**paradigm**", a word which has gone into common usage. Paradigms are more than just equations, laws, statements encapsulated in the chapters of a textbook. The paradigm of Newtonian mechanics was not just Newton's laws of motion, it was also the model or picture of the universe as a deterministic clockwork in which the fundamental properties of things were their position and momentum from which all the rest of their behavior could eventually be derived when Newtonian science was completed. The Newtonian paradigm also included the standard set of apparatus or lab equipment whose behavior was explained, predicted and certified by Newton's laws, and with it a certain strategy of problem-solving. The Newtonian paradigm includes a methodology, a philosophy of science, indeed an entire metaphysics. In his later writing Kuhn placed more emphasis on the role of the **exemplar** – the apparatus, the practice, the impedimenta – of the paradigm than on any verbal expression of its content. The exemplar more than anything defines the paradigm.

Paradigms drive normal science, and normal science is in a crucial way quite different from the account of it advanced by empiricist philosophers of science. Instead of following where data, observation and experiment lead, normal science dictates the direction of scientific progress by determining what counts as an experiment that provides data we should treat as relevant, and when observations need to be corrected to count as data. During normal science, research focuses on pushing back the frontiers of knowledge by applying the paradigm to the explanation and prediction of data. What it cannot explain is outside of its intended domain, and within its domain what it cannot predict is either plain old experimental error or the clumsy misapplication of the paradigm's rules by a scientist who has not fully understood the paradigm.

Under the auspices of normal science, three sorts of empirical inquiries flourish: those which involve redetermining of previously established observational claims to greater degrees of precision, certifying the claims of the current paradigm against its predecessor; the establishment of facts without significance or importance for themselves but which vindicate the paradigm; and experiments undertaken to solve problems to which the paradigm draws our attention. Failure to accomplish any of these three aims reflects on the scientist attempting them, not the paradigm employed. None of these sorts of inquiry is to be understood on the empiricist model of experience testing theory.

The grandest example of the success of normal science in giving priority of belief to theory over data (and thus undermining empiricism) is found in the story of Newtonian mechanics and the planets of Neptune and Pluto. One of the great successes of Newtonian mechanics in the 1700s was predicting the appearance and reappearance of Halley's comet by enabling astronomers to calculate its orbit. In the nineteenth century, apparent improvements in telescopes enabled astronomers to collect data on the path of Uranus which suggested a path different from that

Newtonian theory predicted. As we have seen in Chapter 5, this apparently falsifying observation discredits the "package" of Newton's laws, along with a large number of auxiliary hypotheses about how telescopes work and what corrections have to be made to derive data from observations using them, as well as assumptions about the number and mass of the known planets whose forces act upon Uranus. The centrality of the Newtonian paradigm to normal science in physics did not in fact leave matters underdetermined in the way Chapter 5 suggests. The ruling paradigm dictated that the data on Uranus be treated as a "puzzle", that is, a problem with a "correct" answer to be discovered by the ingenuity of physicists and astronomers applying the paradigm. A physicist's failure to solve the paradigm simply discredited the physicist, not the physics! There could be no question that the theory was wrong; it had to be the instruments, the astronomers, or the assumptions about the number and mass of the planets. And indeed, this was how matters turned out. Accepting the force of the Newtonian paradigm, and the reliability of the instruments which the Newtonian paradigm certified, left only the option of postulating one or more additional planets, as yet undetected (because too small or too distant or both), whose Newtonian gravitational forces would cause Uranus to move in the way the new data suggested. Training their telescopes in the direction from which such forces must be exerted, astronomers eventually discovered first Neptune and then Pluto, thus solving the puzzle set by the Newtonian paradigm. Whereas the empiricist would describe the outcome as an important empirical confirmation of Newton's theory, followers of Kuhn would insist that the paradigm was never in doubt and so neither needed nor secured additional empirical support from the solution to the puzzle.

Normal science is characterized by textbooks, which despite their different authors convey largely the same material, with the same demonstrations, experiments and similar lab manuals. Normal science's textbooks usually contain the same sorts of problems at the back of each chapter. Solving these puzzles in effect teaches scientists how to treat their subsequent research agendas as sets of puzzles. Naturally, some disciplines are, as Kuhn put it, in "pre-paradigm" states, as evinced for example by the lack of textbook uniformity. These disciplines are ones, like many of the social sciences (but not economics), where the lack of commonality among the textbooks reveals the absence of consensus on a paradigm. How the competition in pre-paradigm science gives way to a single winner, which then determines the development of normal science, Kuhn does not tell us. But he does insist paradigms do not triumph by anything like what the experimental method of empiricism suggests. And the reason Kuhn advances is an epistemologically radical claim about the nature of observation in science.

Recall the distinction between observational terms and theoretical terms so important to the project of empiricism. Observational terms are used to describe the data which epistemically control theory, according to the empiricist. The empiricist's problem is that observation seems inadequate to justify the explanatory theories about unobservable events, objects and processes with which science explains the observable regularities we experience in the lab and the world. This problem for empiricism is not a problem for Kuhn, because he denies that there is a vocabulary that describes observations and that is neutral between competing theories. According to Kuhn, paradigms extend their influence not just to theory, philosophy, methodology and instrumentation, but to the lab-bench and the field notebook, dictating observations, not passively receiving them.

Kuhn cited evidence from psychological experiments about optical illusions, gestalt-switches, expectation-effects and the unnoticed theoretical commitments of many apparently observational words we incautiously suppose to be untainted by presuppositions about the world. Consider some examples. Kuhn's example was a red jack of spades and a black jack of hearts which most people don't notice are red and black as they are accustomed to black spades and red hearts. Since Kuhn first made the point, other examples have become common knowledge. In the Mueller-Lyer illusion, two lines of equal length, one with an arrow at each end pointing out, and the other with arrows pointing in, are viewed by western eyes as unequal; but the illusion does not fool people from "non-carpentered societies" without experience of straight lines. The Necker cube, a simple two-dimensional rendering of a transparent cube, is not so identified by those without experience of perspective, and the front–back switch or reversal which we can effect in our perception shows that the act of seeing is not a cognitively innocent one. When Galileo first described the moon as "cratered", his observations already presupposed a minimal theoretical explanation of how the lunar landscape was created – by impacts from other bodies.

Kuhn was not alone in coming to this conclusion. Several opponents of empiricism came in the 1950s to hold this view about observation. They held that the terms in which we describe observations, whether given by ordinary language or scientific neologisms, presuppose divisions or categorizations of the world of experience in ways that reflect prior "theories": the categories we employ to classify things, even categories as apparently theory-free as color, shape, texture, sound, taste, not to mention size, hardness, warmth/coldness, conductivity, transparency, etc., are shot through with interpretation. Instead of seeing a glass of milk, we see "it" as a glass of milk, where the "it" is not something we can describe separately in a theory-neutral vocabulary. Even the words "white", "liquid", "glass", "wet", "cold", or however we seek to describe our sensory data, are as much theory-bound as "magnetic" or "electric" or "radioactive".

Since Kuhn first wrote, this claim that the theoretical/observational distinction is at least unclear and perhaps baseless, has become a lynchpin for non-empiricist philosophy of science. Its impact upon the debate about the nature, extent and justification of scientific knowledge cannot be understated. In particular it makes much more difficult to understand the nature of scientific testing – the most distinctive of science's differences from everything else. Kuhn recognized this consequence, and his way of dealing with it is what made *The Structure of Scientific Revolutions* so influential a work.

A revolution occurs when one paradigm replaces another. As normal science progresses, its puzzles succumb to the application or, in Kuhn's words, "the articulation" of the paradigm. A small number of puzzles continue to be recalcitrant: unexpected phenomena that the paradigm cannot explain, phenomena the paradigm leads us to expect but that don't turn up, discrepancies in the data beyond the margins of error, or major incompatibilities with other paradigms. In each case, there is within normal science a rational explanation for these anomalies; and often enough further work turns an anomaly into a solved puzzle. Revolutions occur when one of these anomalies resists solution long enough, while other anomalies succumb, to produce a crisis. As more and more scientists attach more importance to the problem, the entire discipline's research program begins to be focused around the unsolved anomaly. Initially small numbers of especially younger scientists without heavy investment in the ruling paradigm cast about for a radical solution to the problem the anomaly poses. This will happen usually when a paradigm has become so successful that few interesting puzzles are left to solve. More and more of the younger scientists, especially, with ambitions and names to make, decide to attach more importance to the remaining unsolved puzzle. Sometimes, a scientist will decide that what could reasonably be treated as experimental error is something entirely new and potentially paradigm-wrecking. If the ultimate result is a new paradigm, what the scientist has done is retrospectively labeled a new discovery. When Roentgen first produced X-rays, he treated the result as contamination of photographic plates. The same plates became evidence of a significant phenomenon once paradigm shift had allowed for it. If the ultimate result is not incorporated by a paradigm shift, it gets treated as error – poly-water for example – or worse, fraud – cold-fusion.

In developing a new paradigm, revolutionaries are not behaving in the most demonstrably rational way; nor are their usually elderly establishment opponents who defend the ruling paradigm against their approach, acting irrationally. During these periods of crisis when debate in a discipline begins to focus inordinately on the anomaly, neither side can be said to be acting irrationally. Defenders of the old paradigm have the weight of all its scientific successes to support their commitment. Exponents of the new one have only at most its solution to the anomaly recalcitrant to previous approaches.

Note that during these periods of competition between old and new paradigms, nothing between them can be settled by observation or experiment. This is for several reasons. To begin with, often there is little or no difference between the competing paradigms when it comes to predictive accuracy. Ptolemaic geocentric astronomy with its epicycles was predictively as powerful, and no more mathematically intractable, than its Copernican heliocentric rival. Moreover, "observational" data are already theoretically charged. It does not constitute an unbiased court of last resort. For Kuhn there is in the end no evidentiary court that will decide between competing paradigms which is more rational to embrace, which is closer to the truth, which constitutes scientific progress. This is where the radical impact of Kuhn's doctrine becomes clear.

A persistently unsolved and paradigmatically important anomaly will result in a scientific revolution only when another paradigm appears that can at least absorb the anomaly as a mere puzzle. In the absence of an alternative paradigm, a scientific discipline will continue to embrace its received one. But the grip of the paradigm on scientists is weakened; some among them begin to cast around for new mechanisms, new rules of research, new equipment, and new theories to explain the relevance of the novelties to the discipline. Usually in this "crisis-situation", normal science triumphs; the anomaly turns out to be a puzzle after all, or else it just gets set aside as a problem for the long-term future, when we have more time, money and better research apparatus to throw at it. Revolutions occur when a new paradigm emerges. A new paradigm disagrees radically with its predecessor. Sometimes new paradigms are advanced by scientists who do not realize their incompatibility with ruling ones. For instance, Maxwell supposed that his electromagnetic theory was compatible with the absolute space of Newtonian mechanics, when in fact Einstein showed that electrodynamics requires the relativity of spatiotemporal relations. But the new paradigm must be radically different from its predecessor just insofar as it can treat as a mere puzzle what the previous one found an increasingly embarrassing recalcitrant anomaly. Paradigms are so all-encompassing, and the difference between paradigms is so radical, that Kuhn writes that scientists embracing differing paradigms find themselves literally in different worlds – the Aristotelian world versus the Newtonian one, the Newtonian world versus the quantum-realm. Paradigms are, in Kuhn's words, "incommensurable" with one another. Kuhn took the word from geometry, where it refers to the fact that, for instance, the radius of a circle is not a "rational" fraction of its circumference, but is related to it by the irrational number π. When we calculate the value of π the result is never complete but always leaves a "remainder". Similarly, Kuhn held that paradigms are incommensurable: when one is invoked to explain or explain away another, it always leaves a remainder. But mathematical **incommensurability** is a metaphor. What is this remainder?

According to Kuhn, though a new paradigm may solve the anomaly of its predecessor, it may leave unexplained phenomena that its predecessor successfully dealt with or did not need to deal with. There is a trade-off in giving up the old paradigms for the new, an explanatory loss is incurred at the expense of the gain. For example, Newtonian mechanics cannot explain the mysterious "action at a distance" it required – the fact that gravity exerted its effects instantaneously over infinite distances; this disturbing commitment is something the Aristotelian physics did not have to explain. In effect, "action at a distance" – how gravity is possible – became the anomaly that in part and after two hundred and fifty years or so eventually undid Newtonian mechanics. But explanatory loss is not all there is to incommensurability. For even with some explanatory loss, there might yet be net gain in explanatory scope of the new paradigm. Kuhn suggests that incommensurability is something much stronger than this. He seems to argue that paradigms are incommensurable in the sense of not being translatable one into the other, as poems in one language are untranslatable into another. And this sort of radical incommensurability which makes explanatory loss immeasurable underwrites the further claim that paradigms do not improve on one another, and that therefore science does not cumulate in the direction of successive approximation to the truth. Thus the history of science is like the history of art, literature, religion, politics or culture, a story of changes, but not over the long haul a story of "progress".

Kuhn challenges us to translate seventeenth-century phlogiston chemistry into Lavoisier's theories of oxidation and reduction. It cannot be done, without remainder, without leaving some part of the older theory out, and not necessarily the part of phlogiston theory that was wrong. Perhaps you are inclined to say that phlogiston chemistry was all wrong, and needed to be replaced by a new paradigm. This is the sort of ahistorical approach to the nature of science which Kuhn condemned so strongly. After all, phlogiston chemistry was the best science of its day, it had a long record of success in solving puzzles, organizing instrumentation and securing experimental support. And in the period before the heyday of phlogiston many scientists bent their genius towards alchemy. Isaac Newton was so devoted to the search for how to transmute lead into gold that he may have died of lead poisoning as a result of his many experiments. Are we to say that his mechanics was the greatest scientific achievement of a transcendent genius in physics while his alchemy was the pseudo-scientific mischief of a crackpot? Either we must condemn a century of scientific work as irrational superstition or design a philosophy of science that accepts phlogiston chemistry as science with a capital S. If phlogiston theory is good science, and cannot be incorporated into its successor, it is hard to see how the history of science can be a history of cumulative progress. It seems more a matter of replacement than reduction.

Reduction, recall, is the empiricist's analysis of the interrelation of theories to one another, both synchronically, in the way that chemistry is reducible to physics, and diachronically, in the way that Newtonian seventeenth-century discoveries in mechanics are reducible to the twentieth century's special theory of relativity. But does this reduction really obtain in the way the empiricist supposes. Kuhn explicitly denies that it does. And the reason is incommensurability. Reduction of the laws of one theory to the laws of a more basic theory require that the terms of the two theories share the same meaning. Thus, the notions of space, time and mass should be the same in Newton's theory and in Einstein's special theory of relativity if the latter is just the more general case and the former is the special case, as reduction requires. The derivation of the laws of Newtonian mechanics from those of the special theory of relativity looks simple. All one requires is that "c", the speed of light, travels (like gravity) at infinite speed. The reason one requires this false but simplifying assumption to go from Einstein to Newton is that the special theory of relativity tells us that the mass of an object varies as the ratio of its velocity to that of the speed of light with respect to an observer's frame of reference; Newton's theory tells us, however, that mass is conserved, and independent of relative or absolute velocity whether in proportion to the speed of light or not.

Though the two theories share the same word with the same symbol, **m**, do they share the same concept? Emphatically not. In Newtonian mechanics mass is an absolute, intrinsic, "monadic" property of matter, which can neither be created nor destroyed; it is not a relational property that chunks of matter share with other things, like "is bigger than". In Einstein's theory, mass is a complex "disguised" relation between the magnitude of the speed of light, a chunk of matter and a location or "reference frame" from which the velocity of the chunk is measured; it can be converted to energy (recall $e = mc^2$). The change in the meaning of the word "mass" between these two theories reflects a complete transformation in world-view, a classical "paradigm shift". Once we as historians and philosophers of science see the difference between the meaning of crucial terms in the two theories, and discover that there is no common vocabulary – either observational or theoretical – which they share, the incommensurability between them becomes clearer. But, the physicist is inclined to say, "Look here, the way we teach the special theory of relativity in the textbooks is by first teaching Newton's theory and then showing it's a special case via the Lorenz transformations. It is after all a case of reduction. Einstein was standing on the shoulders of Newton, and special relativity reflects the cumulative progress of science from the special case to the more general one."

To this Kuhn has two replies. First, what is reduced is not Newton's theory, but what we, in the thrall of the post-Newtonian, Einsteinian paradigm imagine is Newton's theory. To prove otherwise requires a translation which would inevitably attribute incompatible properties to mass. Second, it is essential to the success of normal science that once it is up and running, it rewrites the history of previous science to make it appear just another step in the long-term continuous progress of science to cumulative knowledge of everything. The success of normal science requires the disciplining of scientists not to continually challenge the paradigm, but to articulate it in the solution of puzzles. Science would not show the pattern of cumulation which normal science exhibits without this discipline. One way to enforce the discipline of normal science is to rewrite their textbooks to make it appear as much as possible that what went before today's paradigm is part of an inevitable history of progress that leads up to it. Whence the invisibility of previous paradigms, and the empiricist's blindness to what the history of science really teaches. For the empiricist's understanding of science comes from its contemporary textbooks, and their "potted" history.

According to Kuhn we must take seriously the notion that scientific revolutions really are changes of world-view. The crucial shift from Aristotle to Newton was not the discovery of "gravity". It was in part the apparently slight change from viewing the distinction between rest and motion as the difference between zero and non-zero velocity to viewing it as the difference between zero and non-zero acceleration. The Aristotelian sees a body moving at constant velocity as under the influence of a force, "impetus" they called it. The Newtonian sees the body as being at rest, under the influence of no (net) forces. The Aristotelian sees the swinging pendulum bob as struggling against constraining forces. The Newtonian sees the pendulum as in equilibrium, at rest. There is no way to express the notion of "impetus" in Newton's theory, just as there is no way to express Einsteinian mass in Newton's theory. More broadly, Aristotelian science views the universe as one in which things have purposes, functions, roles to play; Newtonian mechanics bans all such "teleological", goal-directed processes in favor of the interaction of mindless particles whose position and momentum at any time together with the laws of nature determine their position and momentum at all other times.

Because a new paradigm is literally a change in world-view, and at least figuratively a change in the world in which the scientist lives, it is often too great a shift for well-established scientists. These scientists, wedded to the old paradigm, will not just resist the shift to the new one, they will be unable to make the shift; what is more, their refusal will be rationally defensible. Or at any rate, arguments against their view will be question-begging because they will presume a new paradigm they do not accept. To some extent we have already recognized the difficulty of falsifying a theory, owing to the underdetermination problem discussed in Chapter 5. Because paradigms encompass much more than theories, it is relatively easy to accommodate what some might call falsifying experience when adjustments can be made not just in auxiliary hypotheses but across a vast range of the intellectual commitments that constitute a paradigm. What is more, there is, recall, no neutral ground on which competing paradigms can be compared. Even if underdetermination of theory by evidence were not a problem, the observational findings on which empiricists admit differing theories may agree, are missing. When allegiance is transferred from one paradigm to another, the process is more like a religious conversion than a rational belief shift supported by relevant evidence. Old paradigms fade away as their exponents die off, leaving the proponents of the new paradigm in command of the field.

Progress is to be found in science, according to Kuhn, but like progress in evolution, it is always a matter of increasingly local adaptation. The Darwinian theory of natural selection tells us that over generations the random variations in traits are continuously filtered by the environment so as to produce an increasing spread of increasingly adaptative variations across a species. But environments change, and one environment's adaptation – say, white coats in the arctic – is another environment's maladaptation – white coats in the temperate forest. So it is with science. During periods of normal science, there is progress as more and more puzzles succumb to solution. But revolutionary periods in science are like changes in the environment, which completely restructure the adaptive problems a paradigm must solve. In this respect, science shows the same sort of progress as other intellectual disciplines show. And this is not surprising, for among the morals many draw from *The Structure of Scientific Revolutions* has been the conclusion that science is pretty much like other disciplines, and can make no claims to epistemic superiority. Rather, we should view the succession of paradigms in the way we view changes in fashion in literature, music, art and culture broadly. We should view competing paradigms the way we view alternative normative ideologies or political movements. When we come to assess the merits of these units of culture, progress in approximating to the truth is rarely at issue. So too for science: in one of the last pages of his book, Kuhn writes, "We may, to be more precise, have to relinquish the notion, explicit or implicit, that changes of paradigm carry scientists and those who learn from them closer and closer to the truth" (The Structure of Scientific Revolutions, 1st edition, ch. 13, p. 170).

2 No place for first philosophy?

The Structure of Scientific Revolutions was published in 1962. The impact of its doctrines within and beyond the philosophy of science is difficult to overstate. Kuhn's doctrine became the lever with which historians, psychologists, sociologists, dissenting philosophers, scientists, politicians, humanists of every strip, sought to undermine the claims of science to objective knowledge, its claims to greater credence than alternative claims about the world. Meanwhile, within philosophy of science, developments that began earlier in the 1950s were reinforcing Kuhn's impact. These developments owe a great deal to the work of a philosopher, W. V. O. Quine, whose thought provided some of the philosophical foundations often held to support Kuhn's historical conclusions.

The traditional objectives of the philosophy of science were to justify science's claims to objective knowledge and to explain its record of empirical success. The explanatory project of the philosophy of science is to identify the distinctive methods that the sciences share which enables them to secure knowledge; the justificatory project consisted in showing that this method is the right one, providing its foundations in logic – both inductive and deductive – and epistemology – whether empiricist, rationalist or some third alternative. These ongoing projects came up against traditional philosophical problems. In particular the underdetermination of theoretical knowledge by observational knowledge has made both the explanatory task and the justificatory one far more difficult. If observations underdetermine theory, then discovering the actual inference rules – the methods – that in fact are employed by science is a complicated matter that will require more than armchair logical theorizing. Philosophy will have to surrender exclusive domain over the explanatory task, if it ever had such domain, to psychologists, historians and others equipped empirically to explore the cognitive processes that take scientists from hypotheses to data and back to theory. More radical has been the effect of underdetermination on the justificatory program. Underdetermination of theory by data means that no single hypothesis is supported or disconfirmed by any amount of observation. If data support theory at all they do so in larger units than the single hypothesis. So it was that empiricist philosophers of science were driven to a "**holism**" about justification: the unit of empirical support is the entire theory – both the hypothesis directly under test, every other part of the theory that supports the tested hypothesis, and all the auxiliary hypotheses needed to deliver the test.

Even more radically, the traditional philosophical gulf between justification and explanation came to be challenged by philosophers themselves. Explanations, as we noted in Chapter 2, cite causes, and causal claims are contingent, not necessary truths. The world could have been otherwise arranged and the laws of nature might have been different. That is why we need to undertake factual inquiry, not logical analysis, to uncover causes and provide explanations. Justification is, however, not a causal but a logical relationship between things. What may cause you to believe something does not thereby constitute evidence that supports your belief as well justified. Observing one thing happen may cause you to believe something, but it won't justify that belief unless there is the right sort of logical relation between the things observed. These logical relations are studied naturally enough by philosophers, who seek their grounds: what makes the rules of logic – deductive or inductive – the right rules for justifying conclusions derived from premises, i.e. from evidence. The traditional philosophical answer to the question what makes these the right rules is that they are necessary truths that could not be otherwise.

Empiricists have a difficulty with this answer because they hold that knowledge is justified by experience and that experience cannot demonstrate necessity. Therefore, logical principles which are to justify reasoning were at risk of being ungrounded themselves. For at least two hundred years the empiricist's solution to the problem is to treat all necessary truths, whether in logic or mathematics, as true by definition, as reports about the meaning of words, conventions we adopt to communicate. As such these statements are true by stipulation. The logical rule which tells us that all inferences of the form

if **p** then **q**

p

therefore

q

is true because it reflects the meanings of the terms "if", "then", "therefore". Similarly, all the truths of mathematics, from 2 + 2 = 4 to the

Pythagorean theorem to Fermat's last theorem (there are no positive integer values of **n** greater than 2 such that $x^n + y^n = z^n$) are simply logically deduced from premises which are themselves definitions.

But twentieth-century work in the foundations of mathematics showed that mathematics cannot simply be composed of definitions and the consequences of them. When it was proved by Kurt Gödel that no set of mathematical statements can be both complete (enabling us to derive all the truths of arithmetic) and consistent (including no contradictions), the empiricist claim that necessary truths were all definitions came undone. Empiricism needed a new theory of necessary truths, or it needed to deny that there are any. This is where holism and underdetermination re-enter the story.

A necessary truth, whether trivially true, like "All bachelors are unmarried" or less obviously true, like "the internal angles of a triangle equal 180 degrees" is one that cannot be disconfirmed by experience. But holism teaches us that the same can be said for statements we consider to be contingent truths about the world, statements like "the spin angular momentum of an electron is quantized" or "the speed of light is the same in all reference frames", or, in the past, Newton's laws of motion. Scientists always prefer to make adjustments elsewhere rather than give up these statements. If holism is right, we can always preserve statements like these as true "come what may", simply by revising some other part of our system of beliefs about the world. But then, what does the difference between necessary truths and contingent ones we are unwilling to surrender come to? Well, necessary truths are true just in virtue of the meaning of words that express them, and contingent ones are true in virtue of facts about the world. But if two statements are both unrevisable, how can we tell empirically whether one is protected from revision because of meanings and the other because of beliefs about the world? Notice this is an empiricist challenge to an empiricist thesis, or as Quine put it, a "dogma": that we can distinguish truth in virtue of meanings from truth in virtue of facts.

What are meanings? Recall the empiricist theory sketched in Chapter 4, which holds that meanings are ultimately a matter of sensory experience: the meaning of a word is given by definition in terms of some basement level of words that name sensory qualities – colors, shapes, smells, textures, etc. This theory of language resonates with our pre-philosophical belief that words name images or ideas in the head. But as we have seen, it cannot make sense of the meaning of many terms in theoretical science. What is more, it is hard to see how we could empirically tell the difference between a truth about sensations which defines a term, and a sentence that reports a fact about the world: suppose we define salty thus: "salty is the taste one gets under standard conditions from sea water". What is the difference between this sentence and "salty is the taste one gets under standard conditions from dissolved potassium chloride". One cannot say the former is true in virtue of meaning, because it is meaning that we are trying to elucidate empirically by contrasting these two sentences. One cannot say that "potassium chloride" is a theoretical term and that makes the difference, because "sea water" is equally not a label we can pin on a sample of clear liquid by mere visual inspection. We had to add the "standard conditions" clause to both sentences, because without them, they would both be false (an anesthetized tongue won't taste either as salty). But having added the clause, both can be maintained as true, come what may in our experience. In short, the meaning of words is not given by the sensory data we associate with them. Or if it is given by sensory experience, the relation is very complex. The conclusion Quine came to was that the "meanings" were suspect and no self-respecting empiricist philosopher should want to trade in them. A conclusion with wider support in the philosophy of science was "holism about meaning", a doctrine similar to and mutually supportive of the epistemological thesis of holism in the way data tests theory.

If there are no meanings, or no truths of meaning distinct from truths about the world, if theory meets data as a whole, and the meaning of a theory's terms are given by their place or role in a theory, then we have not just a philosophical explanation for underdetermination, but a philosophical foundation for incommensurability as well. Or at least we will if we part company from Quine in one respect. Despite his rejection of the empiricist theories of meaning and of evidence, Quine did not surrender his commitment to an observational language with a special role in adjudicating competing scientific theories.

Given a continuing role for observation, we may not be able to compare theories sentence by sentence for observational support, or to translate the purport of competing theories into statements about what exactly we will observe under mutually agreed-upon circumstances. But we will be able rationally to choose between theories on the basis of their all-around powers to systematize and predict observations. The result for Quine and his followers was a sort of pragmatism that retained for science its claim to objectivity.

However, the implications of Quine's critique of empiricism's theory of meaning and of evidence make for a more radical holism about mathematics, all the empirical sciences and philosophy for that matter. If we cannot distinguish between statements true in virtue of meaning and statements true in virtue of facts about the world, then there is no distinction of kind between the formal sciences, like mathematics, and the empirical sciences, such as physics or biology. Traditionally, mathematics –

geometry, algebra, and logic – were held to be necessary truths. In epistemology empiricists differed from rationalists about our knowledge of
these necessities. Empiricists held them to be truths of meaning without
content; this is why they are necessary, because they reflect our decisions
about how to use the concepts of mathematics. Rationalists held that
these truths were not empty or trivial disguised definitions and their
consequences, but truths which experience could not justify. Rationalism
could not provide in the end a satisfactory account of how we can acquire
such knowledge and so went into eclipse, at least as the basis for a viable
philosophy of mathematics and science. But, to the extent that empiricism
could not draw an empirically well-grounded distinction between truth in
virtue of meaning and truth in virtue of facts about the world, its account
of how we have knowledge of necessary truths collapses. Quine's conclusion is that all statements we take to be true are of one kind, that there is
no grounded distinction between necessary truths and contingent ones.
So, mathematical truths simply turn out to be the most central and relatively unrevisable of our scientific hypotheses.

What goes for mathematics, goes for philosophy too – including metaphysics, epistemology, logic and the study of scientific methodology.
Theories in these compartments of philosophy turn out also to be no
different from theoretical claims in the sciences. A theory of the nature,
extent and justification of knowledge will turn out for Quine to be a
compartment of psychology; metaphysics – the study of the basic categories of nature – will turn out to be continuous with physics and the
other sciences, and its best theory will be the one which, when put
together with what we know from the rest of science, gives us the most
adequate account of the world, judged as a whole by its ability to explain
and predict our observations. Methodology and logic also are inquiries to
be pursued together with, and not as independent foundations for, the rest
of science. Those methods and those logical principles are most well-
supported which are reflected in the pursuit of successful science. Here
the notion of "empirical adequacy" which we met in Chapter 4 is relevant. Quine's criterion for theory choice in philosophy and in science is
empirical adequacy.

Instrumentalists argue for their doctrine from the privileged position
of a prior philosophical theory, adherence to a strict empiricism. Quine
rejects the claim that there is some body of knowledge, say, a philosophy
or an epistemology, which has greater credibility than science, and might
provide a foundation for it. Though he holds that science should aim for
empirical adequacy, he does so because this is the criterion of adequacy
which science sets itself; what is more, unlike the instrumentalist, and like
the scientist, Quine takes the theoretical claims of science about unob-
servables not just literally but as among the most well-founded of our

beliefs, because in the package of our beliefs we call science, these are among the most central, secure and relatively unrevisable. In fact, for Quine and his followers, science is as much a guide to philosophy as philosophy is to science. The difference between science and philosophy is one of degree of generality and abstractness, not a difference between necessary truths and factually contingent ones.

The resulting philosophy of science has come to be called "**naturalism**". Among philosophers, naturalism became the successor to empiricism largely as a result of Quine's influence. The label, "naturalist" is one many philosophers of science subsequently adopted despite differences among their philosophies of science. But as Quine defended it, naturalism's chief tenets are, first, the rejection of philosophy as the foundation for science, the arbiter of its methods, or the determinant of its nature and limits; second, the relevance of science to the solution of philosophical problems; third, the special credibility of physics as among the most secure and well-founded portion of human knowledge; and, fourth, the relevance of certain scientific theories as of particular importance to advancing our philosophical understanding, in particular the Darwinian theory of natural selection. The importance of Darwinian theory as a scientific guide to the solution of philosophical problems is owing to its account of how blind mechanistic processes can give rise to the *appearance* to us of purpose and design in a world of blind variation and natural selection. Recall the problem of teleological or goal-directed processes and their causal explanation discussed in Chapter 2. Physical science has no conceptual room for final causes, for effects in the future bringing about causes in the past. Still less does it have scope for an omnipotent designer who brings things about to suit his or her desires. This is why the physical world-view finds so attractive a theory like Darwin's, which provided a causal mechanism – the perpetual occurrence of variation (through mutation and recombination) in traits that just happened to be heritable, and the long-term winnowing out by the environment of those variations that work worse than others. If we can use the same mechanism of random heritable variation and selection by the environment to explain other apparently purposive non-physical processes, especially human affairs, we will have accommodated these processes at least in principle to a single scientifically coherent world-view – a naturalistic philosophy.

Exploiting Darwinism, philosophers have sought to provide a naturalistic account of scientific change, similar in some respects to Kuhn's account of scientific progress as local adaptation. Others have sought an epistemology or an account of how scientists actually reason and theorize as random variation (i.e. creative theoretical speculation) and selection by the environment (i.e. experiment and observation). Others have sought an account of the nature of thought in general by appeal to Darwinian processes. Still other philosophers have made common cause with social scientists in building theories of human behavior from a Darwinian basis. Applying Darwinian theory as a research program in philosophy has expanded widely from Quine's original articulation. Doing so makes concrete naturalism's claim that science and philosophy are of a piece and that our most well-established scientific claims should have as much influence on the framing of philosophical theories as our philosophy may have on science.

But naturalism leaves as yet unsolved a major problem. Recall the distinction between justification and causation. Justification gives grounds for the truth of belief; causation does not. Or at least so it seems. In the empiricist's hands, justification is a logical relation (employing deductive or inductive logic) between evidence (sensory experience) and conclusion, and logic is a matter of meanings. Naturalists, or at least Quineans, cannot help themselves to this way of drawing the distinction between causation and justification. Yet draw it they must. Without recourse to a "first philosophy", some body of *a priori* truths, or even definitions, naturalism can only appeal to the sciences themselves to understand the inference rules, methods of reasoning, methodologies of inquiry and principles of epistemology which will distinguish between those conclusions justified by evidence and those not justified by it.

Now, suppose one asks of a principle of logic, or a methodology, whether this method or rule which justifies conclusions is itself justified or well-grounded. The empiricist has an answer to this question: the rule or method is necessarily true, and its necessity rests on our decision about how to use language. We may dispute this argument, and naturalists will do so, because it trades on notions in dispute between empiricists and naturalists – notions like "necessity" and "meaning". But what can naturalists say when asked to ground their own justificatory rules and methods. Appeal to a "first philosophy", an epistemology prior to and more secure than science, is out of the question. And naturalism cannot appeal to science or its success to ground its rules. For the appeal to a "first philosophy" would be circular, and grounding its rules on science's technological success would be to surrender naturalism to a first philosophy – in this case, one called "pragmatism".

Naturalism justifies the epistemology, logic and methodology it recommends because this trio of theories and rules emerges from successful science – i.e. research programs which provide knowledge – justified conclusions – about the way the world works. But if asked why they claim that successful science provides such justified conclusions, naturalists cannot then go on to cite the fact that successful science proceeds by rules and methods which certify its conclusions as justified, because these rules and methods are themselves certified by science's success. Naturalism would be reasoning in a circle. This is a particularly acute problem for Quine, because many of his arguments against empiricism's answers to these questions, by appeal to concepts of logical necessity and meaning, accused these answers of circular reasoning.

To appeal to the practical, technological, applied success of science might solve the naturalist's justificatory problem. But the result would no longer be naturalism. Science does in fact have a magnificent track record of technological application with practical, pragmatic success. But why should this provide a justification for its claims to constitute knowledge or its methods to count as an epistemology? It does so only if we erect a prior first philosophy. Call it pragmatism, after the early twentieth-century American philosophers – William James, C. S. Peirce and John Dewey – who explicitly adopted this view. This philosophy may have much to recommend it, but it is not naturalism, for it begins with a philo-sophical commitment prior to science, and may have to surrender those parts of science incompatible with it.

Naturalism is thus left with an as yet unfulfilled obligation. It aims to underwrite the objectivity of science, its status as ever-improving knowl-edge of the nature of things. It also aims to reflect the actual character of science in its philosophy of science, without giving either philosophy or history a privileged role in the foundations of science or the under-standing of its claims about the world. But it needs to answer in a way consistent with its own principles, and its critique of competing concep-tions, the question of its own justification.

Summary

According to Kuhn, the unit of scientific thought and action is the para-digm, not the theory. Specifying what a paradigm is may be difficult, for it includes not just textbook presentations of theory, but exemplary problem-solutions, standard equipment, a methodology, and usually even a philosophy. Among the important paradigms of the history of science have been the Aristotelian, the Ptolemaic and the Newtonian in physics. Chemistry before Lavoisier, and biology before Darwin, were "pre-para-digm" disciplines, not yet really "scientific", for without the paradigm there is no "normal science" to accumulate information that illuminates the paradigm. The paradigm controls what counts as data relevant to testing hypotheses. There is, Kuhn argued, along with other opponents of empiricism, no observational vocabulary, no court of final authority in experience. Experience comes to us already laden with theory.

Crisis emerges for a paradigm when a puzzle cannot be solved, and begins to be treated like an anomaly. When the anomaly begins to occupy most of the attention of the figures at the research frontier of the discipline, it is ripe for revolution. The revolution consists in a new paradigm that solves the anomaly, but not necessarily while preserving the gains of the previous paradigm. What the old paradigm explained, the new one may fail to explain, or even to recognize. Whence it follows that scientific change – the succession of paradigms – need not be a progressive change in the direction of successive approximation to the truth.

Observation does not control inquiry, rather inquiry is controlled by scientists, articulating the paradigm, enforcing its discipline, assuring their own places in its establishment, except at those crucial moments in the history of science when things become unstuck and a revolution ensues – a revolution which we should understand as more in the nature of a palace coup than the overthrow of an old theory by one rationally certifiable as better or more correct.

This picture of science is hard to take seriously from the perspective of empiricism, historical or logical. It gained currency among historians, sociologists and psychologists, at the same time as, and in part because of, the influence of the philosopher W. V. O. Quine, who unraveled the tapestry of philosophical theories of science as cumulative observational knowledge about the nature of reality.

Quine began by undermining two distinctions: that between statements true as a matter of logic or form, and statements true as a matter of content or empirically observable fact. It may be surprising, but once this distinction, well-known to philosophy since Kant, is surrendered, everything in epistemology and much in the philosophy of science becomes unstuck. The denial of this distinction gives rise to holism about how theory confronts experience, and to the underdetermination which spawns Kuhn's approach to the nature of science. But it also gives rise to a stronger commitment to science, by some philosophers, than even to philosophy, or at least it gives rise to the idea that we must let contemporary science guide our philosophy, instead of seeking science's foundations in philosophy. Philosophers, largely followers of Quine, who have adopted this view label themselves "naturalists", a term unfortunately that others, especially sociologists adopting incompatible views, have also adopted.

Naturally, neither Quine nor other philosophers are prepared to accept Kuhn's apparent subjectivism about science as the correct conclusion to draw from their attack on empiricism. The problem therefore remains of finding a foundation for science as objective knowledge consistent with these arguments. This is the subject of the next chapter.

Questions

1 Which among various approaches to the study of science – philosophy, history, sociology – is the more fundamental? Do these disciplines compete with one another in answer to questions about science?

2 How would a defender of Kuhn respond to the claim that the history of technological progress which science has made possible refutes Kuhn's claim that science is not globally progressive?

3 Kuhn's arguments against the existence of a level of observation free from theory date back to the 1950s. Have subsequent developments in psychology tended to vindicate or undermine his claims?

4 Quine once said, "philosophy of physics is philosophy enough". Give an interpretation of this claim that reflects Quine's claims about the relation between science and philosophy.

5 Is naturalism question-begging? That is, does according the findings of science control over philosophical theorizing rest on mere assertion that science is our best guide to the nature of reality?

Further reading

Every student of the philosophy of science must read T. S. Kuhn, *The Structure of Scientific Revolutions*. Other important works of Kuhn's include *The Essential Tension*, which includes important reflections on the earlier book. An important review of *Structure* is D. Shapere, "Review of *Structure of Scientific Revolutions*". This and other commentaries on Kuhn are reprinted in G. Gutting, *Paradigms and Revolutions*. A *festschrift* for Kuhn containing several important retrospective papers is Horwich, *World Changes: Thomas Kuhn and the Nature of Science*.

P. Feyerabend, *Against Method*, summarizes a series of papers in which the author champions a philosophically informed version of the most radical interpretation of Kuhn's views.

Quine's attack on empiricism emerges in *From a Logical Point of View*, which contains his extremely influential essay, "Two Dogmas of Empiricism". This too is required reading for anyone interested in the philosophy of science. Quine, *Word and Object*, is a later work that deepens the attack on empiricism, and develops the doctrine of underdetermination so influential on Kuhn and others.

Naturalism is expounded and defended in P. Kitcher, *The Advancement of Science*.

CHAPTER 7
The nature of science and the fundamental questions of philosophy

Overview

Kuhn's doctrines have generally been interpreted so as to give rise to relativism – the theory that there are no truths, or at least that nothing can be asserted to be true independent of some point of view, and that disagreements between points of view are irreconcilable. The result of course is to deprive science of a position of strength from which it can defend its findings as more well-justified than those of pseudo-science; it also undermines the claims of the so-called "hard sciences" – physics and chemistry – to greater authority for their findings, methods, standards of argument and explanation, and strictures on theory-construction, than can be claimed by the "soft sciences" and the humanities. Postmodernists and deconstructionists took much support from a radical interpretation of Kuhn's doctrines for the relativism they embraced.

Among sociologists of science especially, a "**strong program**" emerged to argue that the same factors which explain scientific successes must also explain scientific failures, and this deprives facts about the world – as reported in the results of observations and experiments – of their decisive role in explaining the success of science.

These doctrines had a liberating effect on the social and behavioral sciences and other disciplines which had hitherto sought acceptance by aping "scientific methods" but no longer felt the need to do so. The sociological and even more the political focus on science revealed its traditional associations with the middle classes, and with capitalism, its blindness towards the interests of women, and its indifference to minorities.

But in the end the doctrine that science is not a distinctive body of knowledge, one which attains higher standards of objectivity and reliability than other methods, is not sustainable. This conclusion, however, requires that we return to the fundamental problems in epistemology, the philosophy of language and metaphysics in order to see where philosophy went wrong and led the followers of Kuhn to conclusions of such patent preposterousness. It may also require that we attend to the finds of relevant sciences, such as cognitive and perceptual psychology, to discover whether there are theory-free sources of data and hypothesis-formation in our psychological make-up.

1 From philosophy to history to relativism

The interaction of the naturalism that Quine inspired, and the reading of the history of science which Kuhn provided, together have had a profoundly unsettling impact on the philosophy of science. It shook literally centuries of philosophy's confidence that it understood science. This sudden loss of confidence that we know what science is, whether it progresses and how it does so, and what the sources of its claims to objectivity can be, left an intellectual vacuum. It is a vacuum into which many sociologists, psychologists, political theorists, historians and other social scientists were drawn. One result of the heated and highly visible controversy which emerged was to make it apparent that the solution to problems in the philosophy of science requires a re-examination of the most fundamental questions in other compartments of philosophy, including epistemology, metaphysics, the philosophy of language, and even portions of moral and political philosophy.

Kuhn held that paradigms are incommensurable. This means that they cannot be translated into one another, at least not completely and perhaps not at all; incommensurability also implies explanatory losses as well as gains, and no common measuring system to tell when the gains are greater than losses; incommensurability between paradigms reaches down to their observational vocabulary, and deprives us of a paradigm-neutral position from which to assess competing paradigms. The result is a picture of science not as the succession of more and more complete explanations of a wider and deeper range of phenomena, nor even the persistent expansion of predictive power and accuracy over the same range of phenomena. Rather, the history of science is more like the history of fashions, or political regimes, which succeed one another not because of their cognitive merits, but because of shifts in political power and social influence. This conception of the history of science is an invitation to **epistemic relativism**.

Ethical relativism is the claim that which actions are morally right varies from culture to culture and that there is no such thing as objective rightness in morality. Ethical relativism is seen by its proponents as an open-minded and multicultural attitude of tolerance and understanding about ethnic differences. Ethical relativism leads inevitably to skepticism about whether there really is any such thing as absolute moral rightness at all. Epistemic relativism similarly makes knowledge (and therefore truth) relative to a conceptual scheme, a point of view or perspective. It denies that there can be an objective truth about the way the world is, independent of any paradigm, nor consequently any way to compare paradigms for truth, objectivity or epistemic warrant. Kuhn was ambivalent about whether to plead guilty to the charge of epistemic relativism among paradigms.

But the situation may be even more fraught than Kuhn supposed. For there were philosophers and others eager to transform Kuhn's claims about the broadest paradigms that characterize century-long epochs of normal science, into the incommensurability of individual scientific theories even within the ambit of normal science. And Quine's fundamental philosophical arguments gave them the resources to do so. Most influential among these philosophers was Paul A. Feyerabend. Adopting Kuhn's insights about the irreducibility of Aristotelian mechanics to Newton's theory, and Newtonian mechanics to Einstein's, Feyerabend argued that the impossibility of translating the key concepts of impetus into inertia, or absolute mass into relative mass, reflects a barrier to reduction among all theories. The reason is the holism about meaning that Quine's insights spawned. The meaning of a theoretical term is not given by its connection, direct or indirect, to observation, because theory does not meet observation word by word or even sentence by sentence, but only as a whole. So, meanings are theoretical. The meaning of a theoretical term is given by its place in the structure of the theory in which it figures. Change one or more parts of a theory and the result is not an improvement on the same theory, but an altogether new and different one. Why? Because the new theory is not about the same subject-matter as the old theory, since its words have different meanings. "Electron", though it may be an inscription in Bohr's theory, Thomson's theory, Heisenberg's and Schrodinger's, no more means the same thing in each of them as does "cat" mean the same in "pussy cat", "catastrophe", "cool cat", and "cat o' nine tails".

Denying this holistic claim about meanings requires an entire theory of meaning, or at least a reasoned objection to Quine's attack on meanings. When added to the denial of an observational language that could frame statements about data, statements that might enable us to choose between theories, the result is what Feyerabend praised as "methodological anarchy". He called it methodological anarchy because the result is that there is no cognitive basis to choose between theories. In particular, earlier and "well-established" theories have no claim to our adherence above later and less well-established ones. And Feyerabend praised this outcome because he held that such anarchy stimulates scientific originality and creativity. After all, if Newton had been required to advance a theory which could treat Aristotle's as a special case, or had Einstein been required to do so for Newton, just because of the explanatory and predictive successes of Aristotle's or Newton's theory, neither Newton nor Einstein would have produced the great scientific revolutions which bear their names. Just as moral relativists think their insight emancipatory and enlightened, so did Feyerabend think his epistemic relativism a good thing.

Feyerabend and other relativists would stigmatize naturalism from just this perspective. Like Kuhn, and like naturalists for that matter, relativists will agree that an epistemology and a methodology are parts of a paradigm, or in fact components of a theory, although perhaps these components are expressed grammatically in the imperative instead of the indicative. As such, epistemology and methodology don't provide an independent position from which to adjudicate scientific advance, or even the status of a discipline as "Scientific" with a capital "S". These relativists would seize upon the problem of circularity that faces naturalism to substantiate their claim that any particular theory, paradigm or discipline is but one among many "ways of knowing", and that there is no such thing as one of them being correct and the others mistaken. So far as the relativist is concerned, "Anything Goes". This in fact was the title of a book in which Feyerabend most forcefully argued for this view. Instead of a brief biography, Feyerabend provided his astrological chart on the book's dust-jacket. He meant to suggest that astrology was as informative about the author as personal facts about his education, career and previous books might have been.

But if, from the philosophical point of view, anything goes, the question emerges, why has science taken the particular route that it has over time? For the relativists the answer cannot be that the history of science is the history of inquiry "tracking the truth", changing in the direction of a closer and closer approximation to the truth about the world. Indeed, the way the world is, independently of science, can have no role in determining the shape of particular sciences or science in general. That is because there is literally no way the world is, independent of science. We can take this claim either literally or figuratively, as we will see. If the history of science is not explained by the dispassionate study of the way the world is by objective and disinterested scientists, it must, like all the history of all other social institutions, be the outcome of social, political, psychological, economic and other "non-cognitive" factors. So, to understand science, the particular sciences and the nature of scientific change, relativists argue, we must do social science. For example, to learn why Darwin's theory of evolution as gradual selection of locally fitter traits triumphed does not require that we understand the fossil record, still less the sources of variation and environmental filters. It requires that we understand the social and political forces that shaped theory construction and acceptance in the nineteenth century. Once we understand the ideological needs of nineteenth-century *laissez-faire* capitalism to justify relentless competition in which the less fit were ground under and progress was a matter of market competition, the emergence of the Darwinian paradigm should be no surprise. That the history of science should be rewritten by each successive paradigm is now understandable not just because normal science requires ideological discipline, but because political domination requires it as well.

The denial that tracking the truth had a special role in the explanation of scientific change, which it lacks in, say, changes in literature or fashion, led in the 1980s to an important new movement in the sociological study of science, and a concomitant claim by this movement that sociology must displace philosophy as our source for understanding science. The so-called "strong program" in the sociology of science set out to explain both scientific successes and failures on the same basis. Since what distinguishes those scientific developments that are accepted as advances from those rejected (with hindsight) as mistaken cannot be that the former reflect the way the world works and that the latter do not, both must be explained in the same way. The sociologist David Bloor described this as the "symmetry thesis": it leaves no space for any argument that what explains successful scientific theorizing is that it is more rational than unsuccessful theorizing.

These sociologists and other social scientists sought to study the close details of scientific work, and concluded that, like other social products, scientific agreement was "constructed" through "negotiation" between parties whose interests are not exclusively or perhaps even predominantly describing the way the world works. Rather their interests are personal advancement, recognition, material reward, social status, and other benefits which bear no connection to the declared, publicly stated, advertised objectives of science: the disinterested pursuit of truth. In the hands of radical postmodern students of science, the thesis that scientific findings are constructed becomes the claim that the world external to scientific theory, which realists identify as the independent reality that makes scientific claims true or false, is itself a construction without existence independent of the scientists who agree upon their descriptions of it. This "idealism", according to which to exist is nothing more than to be thought of, goes back in philosophy of science to the eighteenth-century philosopher George Berkeley, and certainly has the explicit support of at least some perhaps incautious remarks of Thomas Kuhn: those which suggest that exponents of differing paradigms live in differing worlds. Moreover, the postmodern sociologists held that the unit of scientific belief is not the individual scientist but the community of scientists at work in a particular research program. Rejecting the first-person point of view associated with traditional scientific philosophy since Descartes, these scholars argued that facts in science were constructed through negotiation among interested parties, instead of discovered by individuals, and researchers subject to replication by other individual researchers.

The social character of knowledge will not only explain the enforcement of consensus, it will also explain certain defects and deficiencies in science, ones which mirror the character of western culture as a whole. Thus, some social scientists – for example, feminists and post-colonialism scholars – have sought to explain the character of science as at least in part the product of patriarchal or racialist agendas. Beginning with blatant examples of sexism among scientists – such as constructing human research populations that exclude women or minorities – or racist pursuit of evolutionary research which stigmatizes races as bearing hereditary limitations which cannot be ameliorated by environmental compensation, these social students of science went on to conclude that much of science reflects such limitations, though few scientists, even those whose work produces such baleful consequences, are conscious of them.

Social critics, commentators and humanists have drawn much inspiration from this social study of science, mainly to "dethrone" science from a position of undue and unjustified authority and respect to which western society has accorded it over the half a millennium since the Renaissance. Critics have begun with the evident fact that science and scientific findings have been misused in two ways. First, they have provided more efficient and effective ways of harming people, other organisms and the environment. Second, these critics of science have gone on to note that science has provided unwarranted rationalization for policies that affect such harms – eugenics for example. Even societies which have not blatantly misused science are often guilty of "scientism" – the unwarranted attribution to science of special epistemic authority. There are, according to these critics, other ways of knowing besides the methods science employs. Disciplines stigmatized as pseudo-science, such as astrology or parapsychology; the theories that stand behind alternative "holistic"therapies in medicine, like homeopathy; non-standard cultivation practices, such as playing music to one's houseplants – these are of equal standing. To deny their epistemic status is simply to argue from the blinkered and question-begging perspective of the Newtonian paradigm, a paradigm for that matter now superseded by scientific advances in cosmology and quantum physics for which we have as yet no acceptable philosophical interpretation. Who can say that when the dust settles in these areas, alternative non-Newtonian ways of knowing will not have been vindicated. To the extent that the social study of science deriving from Kuhn has undermined the credentials of traditional natural science, it has made more controversial the public support for the sciences in those countries, especially Great Britain, where the sociology of science has been most visible and intellectually influential. Some physicists have attacked the social studies of science as weakening public support for natural scientific research. Most scientists simply take these views no more seriously than claims to the effect that the earth is flat. The philosophy of science cannot afford so cavalier an attitude.

Less extreme versions of the relativism associated with the social study of science motivate certain philosophies of social science and certain accounts of the nature of knowledge in the humanities. Thus, qualitative social science has come to defend its methods and results against attack from empirical and quantitative social scientists by claiming for itself the status of a separate and incommensurable paradigm. These defenders of qualitative social science go on to the counter-attack, arguing that the empirical, quantitative, experimental paradigm is incapable of dealing with human meaning, significance and interpretation; that these are the essential dimensions along which human action, emotion and value are to be understood; that the natural-science paradigm cannot even accommodate the notion of semantic meaning, let alone human significance; and that the sterility and frustration of much social science is the result of slavishly attempting to implement an inappropriate paradigm from the natural sciences. The inability to surrender this paradigm in the face of anomalies of the sort that should lead to the questioning of normal science is a tribute to the social and cultural power of natural science as a model for all compartments of human knowledge. Nevertheless, it is the wrong model. So these critics of scientism argue.

2 Could the earth really be flat?

For all of Kuhn's insights into the history of science, something has gone seriously wrong in the development of the social studies of science since his time. So, at least, an unbiased observer (or perhaps someone in the grip of a scientistic paradigm) would suppose. Much of the motivation for the attempt to understand natural science stems from an appreciation of its predictive power and explanatory depth, from the desire to identify its methodological secrets so that they can be applied elsewhere (especially in the social and behavioral sciences) with the same theoretical insights and technological results. When an inquiry so motivated concludes that science is just another religion, just one of a wide variety of ways of looking at the world, none of which can claim greater objectivity than the others, then sometime, somewhere, we have taken a wrong turn in our inquiry.

But where? It is simply not enough to turn one's back on Kuhn's insights, nor on the arguments against the pretensions of science mounted on top of them. Many philosophers of science have concluded that Kuhn's historical account of scientific change has been "over-interpreted"; that he did not intend *The Structure of Scientific Revolutions* as a broadside attack on the objectivity of science. In this they had the support of Kuhn, at least while he still lived. It had not been his intention to cast science down from its claims to objectivity, but to enhance our understanding of it as a human enterprise. Similarly, Quine and his philosophical followers could not countenance the misapplication of their doctrine of underdetermination to support the conclusion that current scientific conclusions are not the most reasonable and well-supported conclusions we can draw about the world. But what Kuhn and Quine may have intended cannot decide what their arguments have in fact established or suggested.

What the defender of scientific objectivity, or at least its possibility, must do, is undermine the claims of incommensurability. To do this one must either attack the assimilation of observation to theorizing, or reconcile it with the possibility of testing theories by observation in a non-question-begging manner. And to show how science can make progress over theoretical change that cumulates knowledge, we will have to show how translation between theories can be effected.

One way defenders of objectivity in science have attempted to reconcile the assimilation of observation to theory with its continued role in testing is to draw a distinction between the categories we adopt for classifying particular items – objects, processes, events, phenomena, data – and the particular acts of classification themselves. Differing and even incommensurable categorical frameworks can be reconciled with agreement about actual findings, thereby making objectivity in the recording of data possible. The difference is like that between the letter-box pigeon-holes in a departmental office and the particular pieces of mail that are distributed to these pigeon holes. Adopting a particular set of labels for boxes doesn't prejudge what pieces of mail will come in. Observations are like pieces of mail. Their descriptions are the labels on the classes into which we sort observations. A hypothesis is a claim that members of one category will also fit into another, or always come together with members of another category. There may be agreement on what falls into any category, and thus a way of testing hypotheses, even when the hypotheses are expressed in terms of categories controlled by a theory that is not itself tested by what falls into its categories. It can even turn out that differing categorical schemes will substantially overlap, thus allowing for agreement about data even between differing categorical frameworks. For example, items which the categorical framework of Einstein's theory of special relativity would classify as "having mass" would also be so classified by Newton's theory, notwithstanding the fact that the two theories mean something quite different by "having mass". And of course, we may surrender categorical systems when they no longer work well, that is, when it becomes difficult to use them to file things uniquely, or too complicated to figure out in which boxes they belong, if some significant numbers of boxes go unexpectedly unfilled, or if we can uncover no interesting hypotheses about which boxes get filled at the same time by the same things. Thus, observation can control theory even when its most basic descriptions reflect pre-established theories, even theories we don't recognize as theories, like those embodied in common sense and ordinary language.

But when one thinks about the notion of a categorical scheme and instances which are classified in accordance with it, the conclusion that there is a place for theory-controlling observations here is simply question-begging. To begin with, items don't come with labels that match up with the labels on the categories: samples of gold don't have the word "gold" printed on them. The simplest act of classification requires hypotheses about other categories. Classifying something as gold requires that we invoke the hypothesis that gold dissolves only in aqua regia. This hypothesis presupposes another set of hypotheses which enable us to tell what aqua regia is. And so on, *ad infinitum*. The *ad infinitum* is due to the fact that there is no basement level of words defined directly by experiences, as the historical empiricists held.

Second, how do we tell the difference between hypotheses about correlations between items in our classifications, like "gold is a conductor", and hypotheses like the one about gold and aqua regia that we need to do the classifying. We need to be able to tell the difference between these hypotheses if we are to treat one set as open to objective test, while the other set is not, owing merely to its classificatory role. We can't argue that the classificatory statements are true by definition (gold = whatever dissolves only in aqua regia), and the "gold is a conductor"-hypothesis is a claim about the world. We cannot do this without first having established a way of empirically telling the difference between definitions and factual claims, and doing this requires still another argument against Quine.

Third, categorical schemes are in fact hypotheses about the world, so the whole distinction breaks down. Consider the most successful categorical scheme science has ever established, Mendeleev's Periodic Table of the Elements. It is a successful categorical scheme because it "divides nature at the joints". The differences between the elements it systematizes are given by atomic theory. In the century after Mendeleev advanced his categorical system, discoveries, especially about nuclear structure and electron-shell-filling, explained the relationship between Mendeleev's rows and columns, and showed that it was more than a merely convenient filing system: it was a set of hypotheses about similarities and differences among elements – known and unknown – which required further and deeper explanation.

Fourth, and finally, it is pretty clear, especially in the case of fundamental theories or paradigms, that the disagreements are not about the individual instances and which categories they are to be filed in. Rather, the disagreements are about the definitions of the categories that make these agreements about classifying impossible, and cannot be compromised: compare Aristotle and Newton on what counts as "rest". Differences in classification reflect incommensurabilities that preclude theory-comparison.

Acceding to the assimilation of observation to theory, while distinguishing categories from their instances, will not preserve the objectivity of science. Rather, the defender of scientific objectivity will have to seek out countervailing evidence from the history of science and better psychological theory and data that counter the psychological claims on which the denial of the distinction between observation and theory rests. Such evidence might show that all humans have some common inherited sensory categorical scheme shaped by evolution to be adapted to success at science or some other enterprise which science can make use of. This is certainly one approach which has been adopted, especially by naturalists. It is open to the question-begging objection of course: appealing to findings and theories in psychology is itself to adopt a non-observational and therefore non-objective basis from which to criticize opposition to objectivity. But then, this is the same kind of evidence which Kuhn and his followers originally cited to undermine the observational theoretical distinction.

Such opponents of objectivity cannot have it both ways. Indeed, one might even charge them with the deepest form of incoherence, for they purport to offer arguments against the objectivity of science. Why should we believe these arguments? Do they constitute an objective basis for their conclusions? What makes their arguments and evidence probative, when the arguments of their opponents are always question-begging? These rhetorical questions do not carry the debate very far. This is largely because opponents of scientific objectivity have little interest in convincing others that their view is correct. Their dialectic position is largely defensive; their aim is to protect areas of intellectual life from the hegemony of natural science. To do so, they need only challenge its pretensions to exclusivity as a "way of knowing". These opponents of scientific objectivity cannot and need not argue for a thesis stronger than epistemic relativism.

The opponent of scientific objectivity's strongest card therefore is the incommensurability of meanings that insulates paradigms and theories even from intertranslation. Incommensurability means that no critique of any theory from the perspective of another is even intelligible. Again, it is not enough to call this doctrine self-refuting, on the ground that in order to communicate it to someone with whom prior agreement has not been established, the doctrine must be false. Such a *reductio ad absurdum* argument is a matter of indifference to opponents of objectivity in science interested not in convincing others but in defending their own view as invincible.

One apparently attractive alternative to the *reductio* argument begins by drawing attention to a fundamental distinction in the philosophy of language: meaning versus reference. Meanings, all will admit, are a great difficulty for philosophy, psychology, linguistics; but reference, or denotation, or extension of a term, seems less problematical. What a word names, what it refers to, is something out there in the world, by contrast with what it means, which may be in the head of a speaker and/or a listener, or for that matter may be a social rule or convention, or a matter of use, or as Quine and his followers might have it, nothing at all. And because the reference of a term is something out there, as opposed to in here (pointing to the head), speakers may agree on what a term names without agreeing on what the term means. Or, in the case of terms that name properties instead of things, like "red" or "loud", we can agree on the instances of things and events that bear these properties. The things which are instances of "red" or "sweet" or "rigid" are members of the "extension" of the term "red" or "sweet" or "rigid". We can agree by inspection on whether things are in the extension of "red" or not, even when we can't get into one another's heads to find out whether what looks red to you looks red to me. We can agree that "Superman" names the same item as "Clark Kent" without concurring that the two expressions have the same meaning (indeed, proper names, like "Clark Kent", have no meaning). Reference and extension, it may be held, are more basic and more indispensable to language than is meaning. Moreover, it is tempting to argue, in the manner of the empiricists of the eighteenth century, that language cannot be learned unless it starts with terms that have only reference or extension or something like it. For if every term has meaning – given by other words – it will be impossible for a child to break into the circle of meaningful terms. To break into language, some words must come to us as understandable solely by learning what they refer to, or at least what events stimulate others to use them.

Finally, there are good arguments to suggest that what is really indispensable for science and mathematics is not that the meanings of terms be given, but that their references be fixed. Take any truth of arithmetic, for example, and substitute any term within it that preserves reference, and the statement will remain true. For example: $3^2 = 9$ remains true when it is expressed as the square of the number of ships in Columbus's 1492 fleet equals the number of fielders on a baseball diamond. If two scientists can agree on the reference of terms, or on the set of things a scientific term is true of – for example, the set of things that have mass, whether Einsteinian or Newtonian – they need not agree on the meaning of the term, or whether a translation is available from one meaning for the term to another. Could agreement on reference be enough to ensure commensurability between scientific hypotheses, theories or paradigms? So some defenders of objectivity, following Israel Scheffler, have argued.

Suppose inquirers could agree on the reference or extension of a set of terms "F" and "G" without even discussing their meanings. Suppose further that this agreement led them to agree on when the extensions of these terms overlap, or indeed are identical. In the latter case, they would have agreed that all Fs are Gs, even without knowing the meanings of "F" or "G". Such meaning-free agreement could be the basis for comparing the differing theories inquirers may embrace, even when these theories are incommensurable. A set of hypotheses about the correlations among objects named by categories on whose reference scientists agree would provide exactly the sort of theory-free court of final authority which would enable us to compare competing and incommensurable theories. Each hypothesis on which scientists concur under their purely referential construal, would be given different meaning by one or another incommensurable theory. But it would be an objective matter of mathematical or logical fact whether, thus interpreted, the hypotheses would be derivable from the theories to be compared. That theory would be best supported which deductively implied those hypotheses on the extension of whose terms there was agreement.

It doesn't take much thought to realize that the only hypotheses which will qualify as purely referential will be ones about objects on which agreement of reference can be established non-linguistically, i.e. by pointing or otherwise picking out things and properties without words. But the only candidates for such hypotheses will be those expressed in the vocabulary of everyday observations! In other words, the appeal to reference is but a covert way of bringing back into play the distinction between observational and theoretical vocabulary that started our problem. One way to see this is to consider how we establish the reference of a term. Suppose you wish to draw the attention of a non-English speaker to an object on your desk, say an apple. You could say "apple", but to a non-English speaker that will not discriminate the apple from anything else on your desk. Suppose you say "that" or "this", while pointing or touching the apple. Well, that will probably work, but it is because your interlocutor knows what an apple is and has a word for it. Now, suppose you wish to draw your interlocutor's attention to the stem of the apple, or the soft brown spot under the stem, or the worm wriggling out of the soft spot, or the depression just under the stem. How might you go about it? What you do now is just about what you did the first time: you point and say the words. And that reveals the problem of working with reference alone. There is no way to tell what you are referring to when you say "this" and point. It could be the apple, the soft spot, the darkest part of the soft spot, the stem, the space occupied by the apple, or any of a large number of other things in the general vicinity of your index finger. Of course this is not a problem when we have other descriptive terms to individuate the particular thing to which we are in fact referring. But the reason this works is of course that these other words have meaning and we know what their meanings are! In short, without a background of meanings already agreed to, reference doesn't work. Pure reference is a will-o'-the wisp. And the guide to reference is in fact meaning. The only purely referential terms in any language are the demonstrative pronouns – "this", "that" – and these fail to secure unique reference. Elsewhere in language the relation between reference and meaning is exactly the opposite of what we need. Securing reference relies on meaning. This is particularly apparent for scientific vocabulary, which is used to refer to unobservable things, processes and events, and their only indirectly detectable properties.

If meaning is our only guide to reference, and the meanings of each of the terms of a theory are given by the role which the terms play in the theory, then theoretical holism about meaning makes reference part of the problem for the defender of scientific objectivity, not part of the solution. If theories or paradigms come complete with categorical systems into which particular objects are classified, then exponents of two different paradigms or theories will not be able to agree on how particular things are classified except by the lights of their respective theories as a whole. This makes each of the theories recalcitrant to any experimental evidence that might disconfirm them. For in classifying events, things, processes, the entire theory is involved, and the description of a counter-example to the theory would simply be self-contradictory. Imagine, given the meaning of the word "rest" in Aristotle's physics, the idea that an object could be moving in a straight line at constant non-zero velocity and have no forces acting up on it? Movement for Aristotle is *ipso facto* not rest, and requires a continually acting force. Nothing would count as being free from the influence of forces which was moving at all. Similarly, whatever it is that an Einsteinian might treat as disconfirming Newton's principle of the conservation of mass, it cannot be anything that a Newtonian could even treat as having mass.

But suppose there is a way adequately to draw the distinction between observation and theorizing, and that we can establish at least in principle the possibility of translating across scientific theories and paradigms. Doing this will only put us in a position to take seriously the problem of underdetermination. For the underdetermination of theory by data in fact presupposes both the observational/theoretical distinction and the comparability of competing theories. Quine certainly did not claim the universality of underdetermination in order to undermine the objectivity of science, only our complacency about what its objectivity consists in. But historians, sociologists and radical interpreters of Kuhn's theory certainly have claimed that underdetermination means that, in science, theory choice is either not rational, or rational only relative to some social, psychological, political or other perspective.

Defenders of the objectivity of science need to show that scientific changes are in fact rational, and not just relative to a point of view. They need to show that the changes in a theory which new data provoke are not just arbitrary, that the acceptance of a new paradigm is not simply a conversion experience, but is justified even by the lights of the superseded paradigm. To do this the philosopher of science must perforce become a historian of science. The philosopher must scrutinize the historical record with at least the care of a Kuhn, to show that beneath the appearances of "madness" which Kuhn and his successor historians catalogued there is a reality of "method". That is, philosophers need to extract from the historical record the principles of reasoning, inference and argument that participants in paradigm shifts and theoretical change actually employed, and then to consider whether these principles can be vindicated as objectivity-preserving ones. This is a task which naturalistic philosophers in particular have set for themselves. They have begun to wrestle with the archives, lab notebooks, correspondence and papers of the scientists engaged in scientific revolutions, great and small, and at the same time kept an eye to what the sciences, especially cognitive science, can tell us about reasoning processes characteristic of humans and the adaptive significance of reasoning for our ability to survive and thrive. As noted above, however, naturalists must at the same time take seriously the charge of begging the question which dogs the attempt to preserve objectivity in the face of the holism of meanings and the want of a clear observational/theoretical distinction.

This charge of question-begging is central to the ways in which opponents of scientific objectivity, progress and cumulation would argue. They would hold that attempts to underwrite the traditional claims of science are not just paradigm-bound, but can be undermined by the very philosophical standards of argument and the substantive philosophical doctrines that defenders of objectivity embrace. If correct, this situation provides a major challenge to those who seek to both understand the nature of science and vindicate its traditional claims. The challenge is nothing less than that which faces philosophy as a whole: to articulate and defend an adequate epistemology, and philosophy of language. And then to show that episodes in the history of the sciences sustain these accounts of what constitutes knowledge and how reference can be secured to the same objects in the world by scientists with profoundly different beliefs about the world. If the philosophy of science has learned one lesson from Thomas Kuhn it is that it cannot let the analysis of what actually happened in science fall exclusively into the hands of those with a relativistic or skeptical agenda.

Some scientists and exponents of "scientism" will be tempted to turn their backs on these issues. They may well suppose that if people who can't or won't do the hard work to understand science wish to pretend it isn't the best approximation we have to the truth about the world, this is their problem. And if there are people whose wish that there be a reality – religious, spiritual, holistic, metaphysical – that transcends anything the science can know about, leads them to the thought that science is blinkered and partial in its account of the truth, well, who are we scientists to wake them from their dogmatic slumbers? But the stakes for science and for civilization are too high simply to treat those who deny its objectivity in the way we would treat those who claim the earth is flat.

Summary

Sociologists, and others eager to reduce the baleful influence of a blinkered, narrow-minded, patriarchal, capitalist and probably racialist paradigm associated especially with Newtonian science, have adopted Kuhn's view of science as a version of epistemological relativism.

Relativism in epistemology, as in ethics, allows for the possibility of alternative and conflicting views without adjudicating which is objectively correct: none are, or rather each is correct from the perspective of some epistemic point of view, and all points of view have equal standing. So far as the strongest sociological interpretation of Kuhn was concerned, science is moved by social forces, not epistemic considerations. Science is a social institution, like any other; and this is how it is to be approached if we wish to understand it.

If the empiricist criticizes this argument as incoherent, the relativist is indifferent. All the relativist requires is an argument that convinces relativism, not one that is even intelligible to, let alone accepted by, the empiricist. But this is the end of all debate, and in recent years many of the most radical of sociologists of science have given up this degree of relativism.

As is evident from a survey of obvious moves in the attempt to restore the fortunes of an empiricist theory of knowledge and metaphysics as well as an empiricist account of language, easy solutions will not avail, and there is still much work to be done by philosophy if we are to understand fully the nature of science. Our project must include an understanding of categorization and observation, both philosophically and psychologically. We must clarify the relations between meaning and reference, and develop an epistemology adequate to deal with underdetermination or to show that it does not obtain, and the philosophy of science must come more fully to grips with the history of science. These are all tasks for a naturalistic philosophy.

Questions

1 According to Kuhn, to be successful, normal science must be authoritarian. Why does Kuhn make this claim, and does it constitute a moral deficiency of science?
2 Defend or criticize: "Now at last we can see that science is just another religion."
3 Explain why epistemic relativism cannot be asserted to be true. To what degree if any does this limit the force of the doctrine of epistemic relativism.
4 "Poetry is untranslatable. Science is not. Therefore, incommensurability is false." Sketch an argument for this view.
5 Go back to the study questions at the end of Chapter 1 and reconsider your answers to them.

Further reading

The classical text predating Kuhn's influence in the sociology of science is R. K. Merton, *The Sociology of Science*.

Many of the works about Kuhn's books – especially collections of papers – mentioned in the last chapter are of great relevance here. Among the most radical of relativist sociologists of science in the period after 1970 are B. Latour and S. Woolgar, *Laboratory Life*; A. Pickering, *Constructing Quarks*; B. Barnes, *Scientific Knowledge and Social Theory*; and D. Bloor, *Knowledge and Social Imagery*. Bloor and Barnes significantly qualified their views 20 years later in B. Barnes, D. Bloor and J. Henry, *Scientific Knowledge: A Sociological Analysis*.

Important work in the philosophy of science sympathetic to the sociological approach is due to H. Longino, *Science as Social Knowledge: Values and Objectivity in Scientific Inquiry*. Longino has also made contributions to feminist philosophy of science.

A defense of classical empiricist theories of knowledge and language and of a realist metaphysics for science along the lines developed in this chapter are to be found in I. Sheffler, *Science and Subjectivity*. Nagel attacks Feyerabend's version of theoretical incommensurability in *Teleology Revisited*, as does P. Achinstein, *The Concepts of Evidence*. L. Laudan, *Progress and its Problems*, develops a problems-based account of the nature of science which seeks to incorporate substantial evidence from the history of science.

Glossary

The introduction of each of these terms is highlighted in **bold** type in the main text.

a priori An *a priori* truth can be known without experience, i.e. its justification does not require knowledge about the way the world is arranged. For example, that 2 is an even number is a statement that can be known *a priori*. Note we may become acquainted with *a priori* truths through experience. But experience is not what justifies them. *A posteriori* is the contradictory of *a priori*. A statement can be known *a posteriori* if and only if its justification is given only by experience.

analytic truth A statement true in virtue of the meanings of the words alone: For example, "all bachelors are unmarried males". Analytic statements can be known *a priori* (see *a priori*). Philosophers following Quine are skeptical that we can distinguish analytic truths from some synthetic truths (see below) by any empirical or behavioral test.

antirealism The denial of scientific realism, according to which it is not reasonable to believe that the unobservable items in the ontology (see below) of any scientific theory actually exist, and that we should adopt an instrumentalist (see below) attitude towards theories which treats them as heuristic devices.

axiomatic system A set of axioms and their logical consequences, as proved by deductive logic. A statement is an axiom in an axiomatic system if it is assumed in the system and not proved. A statement is a theorem in the axiomatic system if it is proved in the system by logical deduction from the axioms. For example, Euclidian geometry begins with five axioms from which all the theorems are derived. The syntactic account of theories (see below) holds that they are axiomatic systems.

Bayesianism An interpretation of probability which holds that probabilities are degrees of belief, or betting odds, purely subjective states of scientists, and that probabilities are not properties of sequences of events in the world. Bayesians employ this conception of probability in order to explain and justify scientists use of data to test hypotheses.

boundary conditions The description of particular facts which are required along with a law to explain a particular event, state or fact, according to the D-N model of explanation. Also known as "initial conditions." For example, in the explanation of the sinking of the *Titanic*, the fact that the ship struck an iceberg of particular size at a particular velocity constitutes the boundary conditions.

causation The relation between events, states, processes in the universe which science sets out to uncover, which its explanations report and which its predictions about provide tests of its explanations. According to the empiricist analysis of causation, following Hume, the causal connection is contingent (see below) and consists in the instancing of regularities, and there is no connection of real necessity between cause and effect. It is widely held that causal sequences differ from accidental sequences, and that counterfactual conditionals (see below) reflect this fact.

ceteris paribus **clause** From the Latin, "other things being equal". A qualification to a generalization that "if P then Q" which reflects the fact that other conditions besides P must obtain for Q to obtain. Thus, striking a match is followed by its lighting, but only *ceteris paribus* for in addition to the striking, oxygen must be present, the match cannot be wet, no strong wind can be blowing, etc.

constructive empiricism The claim, due to van Fraassen that theories are either true or false (realism) but that we cannot tell, and therefore should accept or reject them solely on the basis of their heuristic value in systematizing observations.

contingent truth A statement whose truth is dependent on the way things actually are in nature, and not dependent only on purely logical or other grounds we could know about without experience. Contrast with necessary truth. Example: normal humans have 46 chromosomes (they could have had 48 or 44).

counterexample The identification of one or more items whose existence is incompatible with some statement and therefore a counterexample to its truth. Thus, a particle of finite mass traveling faster than the speed of light is a counterexample to the principle that nothing travels faster than light. One counterexample is sufficient to refute a generalization.

counterfactual conditional A statement of the grammatical form, "if P were the case, then Q would be the case", by contrast with an indicative conditional, "If P is the case, then Q is the case". When a counterfactual is true, even though the sentences contained in its antecedent and consequent (the P and Q) are false, then this suggests the two sentences P and Q report facts which are related as cause and effect, or are connected in a law.

covering law model See deductive-nomological model of explanation.

deductive-nomological (D-N) model An explication of the concept of explanation which requires that every explanation be a deductive argument containing at least one law, and be empirically testable.

deductively valid argument An argument in which if the premises are true the conclusion must be true. For example: any argument of the form "if **p** then **q**, **p**, therefore **q**" is valid. The premises of an argument need not be true for the argument to be valid. For example, "All dogs are cats, all cats are bats, therefore all dogs are bats" is valid. Validity is important because it is truth preserving: in a valid argument, if the premises are true (and of course they might not be), then the conclusion is guaranteed to be true.

disposition A trait of something which it exhibits only under certain conditions. Thus, glass has the disposition of being fragile, that is, it breaks when dropped from a certain height to a surface or a certain hardness. Empiricists hold that dispositions obtain only when there are underlying properties that realize them. A glass is fragile even when it is never broken owing to the molecular structure of the material it is composed of. Dispositions without underlying structures that explain them are suspect to empiricists.

empiricism The epistemological thesis that all knowledge of non-analytic truths (see above) is justified by experience.

epistemic relativism The thesis that there are no propositions knowable, except relative to a point of view, and therefore no truths except relative to points of view. The epistemology associated with any one point of view has no grounds from another point of view.

epistemology The division of philosophy which examines the nature, extent and justification of knowledge, also known as "theory of knowledge". The question whether we can have knowledge of unobservable things is an epistemological question. Compare metaphysics.

exemplar A term employed by Kuhn to characterize the standard textbook example of a solution to a puzzle dictated by normal science, or a particular piece of laboratory equipment along with the rules for its correct employment.

explanandum (pl. *explananda*) The statements that describe what is to be explained in an explanation.

explanans (pl. *explanantia*) The statements that an explanation of some fact consist in.

explication (rational reconstruction) The redefinition of a term from ordinary language which provides necessary and sufficient conditions in place of vague and imprecise meanings, and so eliminates ambiguity and the threat of meaninglessness. This method of philosophical analysis was advocated by the logical positivists. For example, the D-N model explicates the ordinary conception of "explanation."

falsification The demonstration that a statement is false by the discovery of a counter-example (see above). Popper held that the aim of science is to falsify hypotheses and to construct new ones to subject to falsification, since verifying scientific laws (see below) is impossible. If statements can only be tested by employing auxiliary hypotheses, strict falsification is impossible, for it is the set of auxiliary hypotheses and the hypothesis under test which is falsified, and not any one particular statement among them.

holism The doctrine that scientific hypotheses do not meet experience for testing one at a time, but only in large sets, so that falsifications do not undermine one particular statement (see falsification) and confirmations do not uniquely support one particular set of statements (see underdetermination).

hypothetico-deductivism The thesis that science proceeds by hypothesizing general statements, deriving observational consequences from them, testing these consequences to indirectly confirm the hypotheses. When a hypothesis is disconfirmed because its predictions for observation are not borne out, the scientist seeks a revised or entirely new hypothesis.

incommensurability The supposed untranslatability of one theory or paradigm into another. If paradigms or theories are incommensurable, then there will be no possibility of reduction (see below) between them, and in moving from one to another, there will be explanatory losses as well as gains.

inductive argument An argument in which the premises support the conclusion without guaranteeing its truth, by contrast to a deductive argument. For instance, that the sun has risen many days in the past is good grounds to believe it will do so tomorrow, but does not make it logically certain that it will.

inductive-statistical (I-S) model of explanation An adaptation of the deductive nomological model to accommodate explanations that employ probabilistic generalizations instead of strict laws. Probabilistic laws do not deductively entail the events they explain, and therefore the model differs sharply from the D-N model.

inference to the best explanation A form of argument employed in science to infer the existence of otherwise not directly observable or detectable mechanisms on the ground that hypothesizing them best explains observations. A similar pattern of reasoning purports to establish scientific realism (see below) on the grounds that only the approximate truth of current scientific theories can explain the technological success of science.

initial conditions See boundary conditions.

instrumentalism The thesis that scientific theories should be treated as heuristic devices, tools for organizing our experiences and making predictions about them, but that their claims about unobservable things, properties, processes and events should not be taken as literally true or false.

logical empiricism Synonym for logical positivism, which reflects the affinity of this school of philosophy to the British empiricists, Locke, Berkeley and Hume.

logical necessity A statement is logically necessary if its truth follows from the laws of logic alone, or if its denial is self-contradictory. For example "two is an even number" is logical necessity.

logical positivism A school of philosophy of the first half of the twentieth century, aiming to combine empiricism and advances in logic to show all outstanding philosophical problems could be shown to be linguistic and solved by analysis or explication (see definition), or rational reconstruction of language. Logical positivists followed empiricists in holding that the only meaningful terms and statements refer to what experience can verify, whence their "verificationist criterion of meaningfulness".

long-run relative frequency An interpretation of probability according to which a statement of the probability of an outcome (say, tails on a coin flip) is equal to the total number of occurrences of the outcome (tails), divided by the total number of trials (all the coin flips), over the "long run", i.e. a run extended indefinitely into the future.

metaphysics The division of philosophy which examines the basic kinds of things there are in the universe. For example, the question "are there unobservable things" is a metaphysical question. Compare epistemology.

model An intentionally simplified description of the regularities governing a natural process or a definition of such a system, usually mathematical and sometimes derived from a more general, less idealized or simplified theory, but sometimes developed independently of any theory. See also semantic approach to theories.

natural kind A metaphysical (see above) concept. By contrast with an artificial kind, a natural kind is a type of state, event, process or thing with existence independent of our classificatory interests. Thus, natural kinds are those which figure in natural laws (see below). "State capital" is an artificial kind, "Acid" is a natural kind.

natural law A regularity that actually governs processes in nature and which science sets out to discover. Laws are usually thought to be of the conditional form, "if a then b" or "all as are bs". Natural laws are often held to be true exceptionalness regularities that underlie causal relations. See scientific law.

naturalism The philosophical thesis that the findings and methods of the natural sciences are the best guides to inquiry in philosophy, and particularly the philosophy of science. Naturalism rejects the claim that philosophy provides *a priori* foundations for science, and instead attempts to solve philosophical problems by exploiting theories in natural science. Naturalists are especially eager to derive insights for philosophy from Darwinian evolutionary theory.

necessary truth A statement whose truth is not dependent on any contingent fact about the way the world just happens to be, but which reflects the only way things could be arranged. Contrast with contingent truth. For example, that 2 is an even number is a necessary truth.

necessity See logical necessity, physical necessity.

normal science The articulation of a paradigm, in which the scientist's task is to apply the paradigm to the solution of puzzles. Failure to solve puzzles is the fault of the scientist not the paradigm. Persistent failure makes a puzzle an anomaly and threatens a revolution which may end the paradigm's hegemony.

normative Having to do with norms about the way things ought to be, as opposed to "positive" or "descriptive", having to do with the way things actually are, thus the realm of values, morality, ethics, policy.

ontology Metaphysics, the study of the basic kinds of things that exist. In the philosophy of science, more narrowly, the ontology of a theory are the kinds of things the theory is commitment to the existence of. Thus, Newtonian mechanics is committed to the existence of mass as an intrinsic property of things. Einsteinian mechanics is committed to mass as a relational property of things and their reference frames.

paradigm A term employed by Kuhn to characterize a scientific tradition, including its theory, textbook problems and solutions, its apparatus, methodology, and its philosophy of science. Paradigms govern normal science (see above). The term has come into general use to describe a world-view.

partial interpretation Compare scientific realism.

physical necessity A statement is physically necessary if it is a law of nature or its truth follows from the laws of nature. Thus, it is physically necessary that no quantity of pure plutonium can have a mass of 100,000 kilograms for the laws of physics tell us that long before it reached this mass, it would explode.

positivism See logical positivism.

pragmatics The study of the contexts of communication which effect the meaning and success of an utterance. It is often held that the deductive nomological model of explanation ignores the pragmatic dimensions along which we measure the success of an explanation requested and provided, in favor of purely non-pragmatic matters of logic and meaning.

prior probability In the Bayesian interpretation of probability, the prior probability is the betting odds assigned to a hypothesis before some new evidence is acquired that may change its probability via Bayes' theorem. According to Bayesianism, a scientist can begin with any assignment of a prior probability. Provided certain conditions obtain, so long as the scientist employs Bayes' theorem, the probabilities assigned to the hypothesis will eventually converge on the correct values.

probabilistic propensity The disposition some item has to exhibit some behavior with a certain frequency. For example, uranium atoms have the probabilistic propensity to emit alpha particles. Such propensities are mysterious because there is no underlying property of the systems which exhibit them that further explains the frequency of the behavior in question. Compare the disposition to be magnet which is explained by the orientation of electrons, or the disposition to be fragile which is explained by chemical structure. Nothing explains a uranium atom's disposition to emit alpha particles with a certain frequency.

probability Either the subjective degree of belief that some proposition is true (Bayesian betting odds, see above) or the long-run relative frequency of something's happening under certain circumstances (weather-report probabilities), or the sheer likelihood that a given event will happen (probabilistic propensities in physics, see above). There are philosophical problems associated with each of these three definitions of probability.

projectable The property of a term or predicate that it names a natural kind (see above) and that the property can figure in natural laws. Coined by Goodman in his treatment of the problem of "grue" and bleen".

realism See scientific realism; antirealism. The term is also employed to describe the position of Plato and his followers that numbers are real through abstract particular objects, and that properties, like being red or redness, exist independent of their instances – particularly red things.

reduction The relation between a less general and a more general theory in the same domain that enables the more general theory to explain the (approximate) truth of the less general theory, usually by logical derivation of the laws of the less general theory from the laws of the more general one. Thus, Newtonian mechanics is said to reduce Kepler's laws of planetary motion. Reduction will not obtain if theories are incommensurable (see above).

scientific law Our best estimate as to a natural law. For example, Newton's inverse square law of gravitational attraction was for a long time held to describe an exceptionalness regularity true everywhere and always, and therefore to constitute a natural law.

scientific realism The thesis that the claims of theoretical science must be treated as literally true or false, and that if we accept a theory as true, we are committed to the existence of its ontology (see above), the things it says there are, even if we cannot detect them. Compare antirealism, instrumentalism.

semantic approach to theories The claim that theories are not axiomatic systems (the syntactic approach, see below), but are sets of models, that is definitions of relatively simple systems with greater or lesser applicability to the world. The semantic approach is neutral on whether the models that constitute a theory reflect some underlying mechanism that explains their applicability.

strong program in the sociology of science The attempt to trace the nature of scientific change without relying on the fact that some theories are true or more approximately true than others. This program is motivated by the idea that since, as Kuhn has shown, there are losses as well as gains in scientific revolutions, and epistemic considerations cannot explain which theories triumph, the explanation of why they do so should appeal to factors no different from the factors which explain why some theories fail.

syntactic approach to theories The claim that theories are axiomatic systems in which empirical generalizations are explained by derivation from theoretical laws.

synthetic truth A statement true at least in part in virtue of contingent facts about the world. Thus, that "there are satellites circling Jupiter" is a synthetic truth. According to empiricism (see above), synthetic truths cannot be known *a priori*.

teleological explanation To explain some fact, event, process, state or thing by identifying the purpose, goal or end which it serves to attain. Since attaining a goal usually comes later, and sometimes does not obtain at all, such explanations do not appear to be causal, and are therefore suspect.

testability A statement is testable if definite consequences for observation can be inferred from it and compared to observations. Logical positivists demanded that all meaningful statements be testable. Post-positivist philosophers have accepted that no single statement is testable by itself.

theory See semantic approach, and syntactic approach.

underdetermination Theory is alleged to be underdetermined by data in that for any body of observational data, even all the observational data, more than one theory can be constructed to systematize, predict and explain that data, so that no one theory's truth is determined by the data.

verification To establish the truth of a claim usually by observation. Positivists embraced a verificationism theory of meaning, according to which a statement was meaningful if and only if it was verifiable.

Bibliography

Achinstein, Peter (1967) *The Concepts of Evidence*, Baltimore, Johns Hopkins University Press.

Achinstein, Peter (1983) *The Nature of Explanation*, Oxford, Oxford University Press.

Achinstein, Peter (1988) "The Illocutionary Theory of Explanation", in Joseph Pitt (ed.), *Theories of Explanation*, Oxford, Oxford University Press.

Allen, C., Bekoff, M., and Lauder, G. (1998) *Nature's Purposes*, Cambridge, Mass., MIT Press.

Ayer, A. J. (1961) "What is a Law of Nature", in *The Concept of a Person*, London, Macmillan.

Barnes, Barry (1974) *Scientific Knowledge and Social Theory*, London, Routledge.

Barnes, Barry, Bloor, David, and Henry, John (1996) *Scientific Knowledge: A Sociological Analysis*, Chicago, University of Chicago Press.

Beauchamp, Tom L., and Rosenberg, Alex (1981) *Hume and the Problem of Causation*, Oxford, Oxford University Press.

Berkeley, George (1710) *Principles of Human Knowledge*.

Bloor, David (1974) *Knowledge and Social Imagery*, London, Routledge.

Boyd, B., Gaspar, P., and Trout, J. D. (1991) *The Philosophy of Science*, Cambridge, Mass., MIT Press.

Braithwaite, Richard B. (1953) *Scientific Explanation*, Cambridge, Cambridge University Press.

Burtt, Edwin A. (1926) *The Metaphysical Foundations of Modern Physical Science*, London, Routledge.

Butterfield, Herbert (1965) *The Origins of Modern Science*, New York, Free Press.

Carnap, Rudolph (1952) *The Continuum of Inductive Methods*, Chicago, University of Chicago Press.

Cartwright, Nancy (1983) *How the Laws of Physics Lie*, Oxford, Oxford University Press.

Churchland, Paul, and Hooker, Clifford (eds) (1985) *Images of Science: Essays on Realism and Empiricism*, Chicago, University of Chicago Press.

Cohen, I. Bernard (1985) *The Birth of a New Physics*, New York, Norton.

Conant, James B. (1957) *Harvard Case Histories in the Experimental Sciences*, Cambridge, Mass., Harvard University Press.

Curd, Martin, and Cover, Jan A. (1997) *Philosophy of Science: The Central Issues*, New York, Norton.

Darwin, Charles (1979) *On the Origin of Species*, New York, Avenel.

Dawkins, Richard (1986) *The Blind Watchmaker*, New York, Norton.

Duhem, Pierre (1954) *The Aim and Structure of Physical Theory*, New York, Doubleday.

Feyerabend, Paul (1975) *Against Method*, London, Verso.

Feynman, R. (1984) *QED: The Strange Story of Light and Matter*, Princeton, Princeton University Press.

Fine, Arthur (1986) "The Natural Ontological Attitude", in *The Shakey Game*, Chicago, University of Chicago Press.

Glymour, Clark (1980) *Theory and Evidence*, Princeton, Princeton University Press.

Goodman, Nelson (1973) (first published 1948) *Fact, Fiction and Forecast*, 3rd edn, Indianapolis, Bobbs-Merrill.

Gutting, Gary (1980) *Paradigms and Revolutions*, Notre Dame, University of Notre Dame Press.

Hempel, Carl G. (1965) *Aspects of Scientific Explanation*, New York, Free Press.

Hempel, Carl G. (1988) "Provisos", in A. Grunbaum and W. Salmon (eds), *The Limitations of Deductivism*, Berkeley, University of Chicago Press.

Hoefer, C., and Rosenberg, A. (1994) "Empirical Equivalence, Underdetermination and Systems of the World", *Philosophy of Science* 61: 592–607.

Horwich, Paul (1982) *Probability and Evidence*, Cambridge, Cambridge University Press.

Horwich, Paul (1993) *World Changes: Thomas Kuhn and the Nature of Science*, Cambridge, Mass., MIT Press.

Hume, David (1888) *A Treatise of Human Nature*, Oxford, Oxford University Press.

Hume, David (1974) *Inquiry Concerning Human Understanding*, Indianapolis, Hackett.

Jeffrey, Richard (1983) *The Logic of Decision*, Chicago, University of Chicago Press.

Kant, Immanuel (1961) *The Critique of Pure Reason*, London, Macmillan.

Kitcher, Philip (1995) *The Advancement of Science*, Oxford, Oxford University Press.

Kneale, William (1950) *Probability and Induction*, Oxford, Oxford University Press.

Kuhn, Thomas S. (1957) *The Copernican Revolution*, Cambridge, Mass., Harvard University Press.

Kuhn, Thomas S. (1977) *The Essential Tension*, Chicago, University of Chicago Press.

Kuhn, Thomas S. (1996) *The Structure of Scientific Revolutions*, 3rd edn, Chicago, University of Chicago Press.

Latour, Bruno, and Woolgar, Steve (1979) *Laboratory Life*, London, Routledge.

Laudan, Larry (1977) *Progress and its Problems*, Berkeley, University of California Press.

Leibniz, G. W. (1981) *New Essays on Human Understanding*, Cambridge, Cambridge University Press.

Leplin, Jarrett (ed.) (1984) *Scientific Realism*, Berkeley, University of California Press.

Leplin, Jarrett (1988) *A Novel Argument for Scientific Realism*, Oxford, Oxford University Press.

Leplin, J., and Laudan, L. (1991) "Empirical Equivalence and Underdetermination", *Journal of Philosophy* 88: 449–72.

Lewis, David (1974) *Counterfactuals*, Oxford, Blackwell.

Lewis, David (1984) "Causation", in *Philosophical Papers*, vol. 2, Oxford, Oxford University Press.

Lloyd, Elizabeth (1987) *The Structure of Evolutionary Theory*, Princeton, Princeton University Press.

Locke, John (1690) *Essay on Human Understanding*.

Longino, Helen (1990) *Science as Social Knowledge: Values and Objectivity in Scientific Inquiry*, Princeton, Princeton University Press.

Mach, Ernst (1906) *The Analysis of Sensation*.

Mackie, John L. (1973) *Truth, Probability and Paradox*, Oxford, Oxford University Press.

Mackie, John L. (1974) *The Cement of the Universe*, Oxford, Oxford University Press.

Merton, Robert K. (1973) *The Sociology of Science*, Chicago, University of Chicago Press.

Mill, John S. (1843) *A System of Logic*.

Miller, Richard (1987) *Fact and Method: Explanation, Confirmation and Reality in the Natural Sciences*, Princeton, Princeton University Press.

Nagel, Ernest (1977) *Teleology Revisited*, New York, Columbia University Press.

Nagel, Ernest (1979) *The Structure of Science*, Indianapolis, Hackett.

Nagel, Ernest, and Newman, James R. (1954) *Gödel's Proof*, New York, State University of New York Press.

Newton-Smith, William (1981) *The Rationality of Science*, London, Routledge.

Pickering, Andrew (1984) *Constructing Quarks*, Chicago, University of Chicago Press.

Pitt, Joseph (ed.) (1988) *Theories of Explanation*, Oxford, Oxford University Press.

Popper, Karl R. (1959) *Logic of Scientific Discovery*, New York, Basic Books. (First published in German, 1935.)

Popper, Karl R. (1984) *Objective Knowledge*, New York, Harper and Row.

Quine, Willard. V. O. (1951) *From a Logical Point of View*, Cambridge, Mass., Harvard University Press.

Quine, Willard. V. O. (1961) *Word and Object*, Cambridge, Mass., MIT Press.

Railton, Peter (1988) "A Deductive-Nomological Model of Probabilistic Explanation", in Joseph Pitt (ed.), *Theories of Explanation*, Oxford, Oxford University Press.

Reichenbach, Hans (1938) *Experience and Prediction*, Chicago, University of Chicago Press.

Reichenbach, Hans (1951) *The Rise of Scientific Philosophy*, Berkeley, University of California Press.

Rosenberg, Alex (1985) *The Structure of Biological Science*, Cambridge, Cambridge University Press.

Rosenberg, Alex (1992) *Philosophy of Social Science*, Boulder, Westview.

Salmon, Wesley (1966) *The Foundations of Scientific Inference*, Pittsburgh, University of Pittsburgh Press.

Salmon, Wesley (1984) *Scientific Explanation and the Causal Structure of the World*, Princeton, Princeton University Press.

Salmon, Wesley (1988) "Statistical Explanation and Causality", in Joseph Pitt (ed.), *Theories of Explanation*, Oxford, Oxford University Press.

Salmon, Wesley (1989) *Four Decades of Scientific Explanation*, in Wesley Salmon and Philip Kitcher, *Scientific Explanation*, Minnesota Studies in the Philosophy of Science 13, Minneapolis, University of Minnesota Press.

Salmon, Wesley, and Kitcher, Philip (1989) *Scientific Explanation*, Minnesota Studies in the Philosophy of Science 13, Minneapolis, University of Minnesota Press.

Savage, Leonard (1972) *Foundations of Statistics*, New York, Dover.

Shapere, Dudley (1964) "Review of *Structure of Scientific Revolutions*", *Philosophical Review* 73: 383–94.

Shapin, Steven (1996) *The Scientific Revolution*, Chicago, University of Chicago Press.

Sheffler, Israel (1976) *Science and Subjectivity*, Indianapolis, Bobbs-Merrill.

Smart, J. J. C. (1968) *Between Science and Philosophy*, London, Routledge.

Spector, Marshall (1968) *Concepts of Reduction in Physical Science*, Philadelphia, Tempel University Press.

Stove, David C. (1967) *Hume, Probability and Induction*, Oxford, Oxford University Press.

Suppe, Fredrick (1977) *The Structure of Scientific Theories*, Urbana, University of Illinois Press.

Thompson, Paul (1989) *The Structure of Biological Theories*, Albany, State University of New York Press.

Tooley, Richard M. (1987) *Causation: A Realist Approach*, Oxford, Oxford University Press.

van Fraassen, Bas (1980) *The Scientific Image*, Oxford, Oxford University Press.

van Fraassen, Bas (1988) "The Pragmatic Theory of Explanation", in Joseph Pitt (ed.), *Theories of Explanation*, Oxford, Oxford University Press.

Westfall, Richard (1977) *The Construction of Modern Science*, Cambridge, Cambridge University Press.

Wright, Larry (1976) *Teleological Explanation*, Berkeley, University of California Press.

Index